Brain Protection

Morphological, Pathophysiological
and Clinical Aspects

Edited by

K. Wiedemann and S. Hoyer

With 55 Figures and 28 Tables

Springer-Verlag
Berlin Heidelberg New York Tokyo 1983

Professor Dr. Klaus Wiedemann
Abteilung für Anaesthesiologie
im Zentrum Chirurgie der Universität Heidelberg
Im Neuenheimer Feld 110, 6900 Heidelberg 1, FRG

Professor Dr. Siegfried Hoyer
Institut für Pathochemie und Allgemeine Neurochemie
im Zentrum Pathologie der Universität Heidelberg
Im Neuenheimer Feld 220–221, 6900 Heidelberg 1, FRG

ISBN-13: 978-3-642-69177-5   e-ISBN-13: 978-3-642-69175-1
DOI: 10.1007/978-3-642-69175-1

Library of Congress Cataloging in Publication Data. Main entry under title: Brain protection. Proceedings of a conference held at Hirschhorn castle, West Germany. Bibliography: p. Includes index. I. Cerebral ischemia – Congresses. 2. Brain damage – Prevention – Congresses. I. Wiedemann, K. (Klaus), 1940– . II. Hoyer, S. (Siegfried), 1933– .   RC388.5.B73 1983 616.8'1 83-6760

© Springer-Verlag Berlin Heidelberg 1983
Softcover reprint of the hardcover 1st edition 1983

The use of registered names, trademarks, etc. in the publication does not imply, even in the absence of a specific statement, that such names are exempt from the relevant protective laws and regulations and therefore free for general use.

Product Liability: The publisher can give no guarantee for information about drug dosage and application thereof contained in this book. In every individual case the respective user must check its accuracy by consulting other pharmaceutical literature.

2125/3140–543210

Preface

Significant progress has doubtlessly been made in the field of cerebral protection compared to earlier centuries, as recently reviewed by Elisabeth Frost (6). She cites the recommendations for treatment of brain trauma by Areteus, a Greek physician of the second century A.D. He expressed quite modern views with regard to the need for prompt action considering complications that follow even minor symptoms. He advised burr holes for evacuation of hematoma in seizures, the use of diuretics and, most interestingly, also hypothermia. German surgeons of the 17th century had little more to offer than prescriptions of which the most effective constituent was alcohol (10). Thus, Sir Astley Cooper was probably the next surgeon to make noteworthy contributions when advising the use of leeches to the temporal artery and other means of bleeding instead of surgical intervention in cases of raised intracranial pressure (loc. cit. 6).

Although our knowledge has greatly expanded during the last two decades, extensive discussions have led to only few conclusions. Promising results from animal studies were translated to clinical situations only to yield controversial and sometimes confusing results. Since the observations of Brierly (5) on ischemic cell damage, improved information on structural aspects, probably even related to concomitant biochemical studies, should allow the validity of therapeutic concepts to be verified. Investigations on cerebral ischemia have led to the differentiation of synaptic transmission failure and membrane failure. This in turn has enabled us to understand more clearly the different effects of barbiturate or lidocaine application or/and hypothermia (3). Expanding knowledge in this field may help to define indications for these procedures and probably to explain their failure in some conditions.

Liberation of free fatty acids has been characterized as a substantial event in cerebral ischemia, and in case of reperfusion, they may form substrates for vasoconstrictive substances or even free radicals, causing vascular or neuronal damage (11).

Catecholamines have been traced as stimulating lipolysis by enhanced activity of cAMP, and therapeutic concepts may evolve from this pathogenic principle (9).

The role of disturbed calcium homeostasis in the pathogenesis of cerebral ischemia has been identified to some extent. Thus, interest is focused on calcium blockers in therapy of ischemic and anoxic brain damage (8). Information on cerebral energy metabolism during ischemia and recovery is still needed, especially when studied under the influence of cerebral metabolic depressants. It would be interesting to differentiate their potential degree of cerebral protection as regards the action on functional metabolism or the diminished load on the sodium-potassium transport system (2). Ideally monitoring of cerebral ischemia should form the basis for therapeutic decisions, and it is clearly needed in proving effectiveness of prophylactic procedures advocated in cerebrovascular surgery for instance.

Clinical trials on brain protection still yield controversial results. Apart from clearly discerning states of focal ischemia, global ischemia and brain trauma and their necessarily different treatments, certainly protection from ischemic damage and its therapy have to be differentiated. Cerebral metabolic depressants commonly employed in anesthesia have been favored in this field. However, barbiturates have been restricted in use partly because of contradictory results in experiments on global ischemia (4,7,12) and partly because of their severe side effects on circulation. In some institutions, they have been replaced by althesin, by etomidate or by gamma-hydroxybutyric acid. Phenytoin has recently attracted interest due to its cerebral metabolic depressant effect as well as its inhibition of potassium efflux in dose-related manner (1). New contributions to this field will hopefully yield more information on rational use of these drugs.

In order to discuss these issues among scientists and clinicians, a conference was held at Hirschhorn castle, West Germany. The proceedings of this conference are presented in this volume.

It cannot be claimed with certainty that firm ground has been reached in even some of these fields, but it is nevertheless hoped that these contributions will make worthwhile reading for those engaged in research and clinical application of cerebral protection.

References

1. Atru AA, Michenfelder JD (1981) Anoxic cerebral potassium accumulation reduced by phenytoin: Mechanisms of cerebral protection? Anesth Analg 60:41–45
2. Astrup J (1982) Energy-requiring cell functions in the ischemic brain. Their critical supply and possible inhibition in protective therapy. J Neurosurg 56:482–497
3. Astrup J, Skovsted P, Gjerris F, Sørensen HR (1981) Increase in extracellular potassium in the brain during circulatory arrest: effects of hypothermia, lidocaine and thiopental. Anesthesiology 55:256–262
4. Bleyaert AL, Nemoto EM, Safar P, Stezoski SW, Mickell JJ, Moossy J, Gutti RR (1978) Thiopental amelioration of brain damage after global ischemia in monkeys. Anesthesiology 49:390–398
5. Brierley JB (1973) Pathology of cerebral ischemia. In: McDowell FH, Brennan RW, (eds.) Cerebrovascular diseases. Grune and Stratton, New York, pp 59–75
6. Frost EAM (1981) Brain Preservation. Anesth Analg 60:821–832

7. Gisvold SE, Safar P, Hendricks H, Alexander H (1981) Thiopental treatment after global brain ischemia in monkeys. Anesthesiology 55:A97
8. Hermans C, De Reese R, Van Loon J, Loots W, Jagenau AHM (1982) A cardiac arrest model in rats for evaluating the antihypoxic action of flunarizine. Europ J Pharmacol 81:137–140
9. Nemoto EM (1978) Pathogenesis of cerebral ischemia-anoxia. Crit Care Med 6:203–214
10. Scultetus DJ (1656) Wund-Artzneyisches Zeughauß. Gerlin, Frankfurt. Reprint: Kohlhammer, Stuttgart 1974
11. Siesjö BK (1981) Cell damage in the brain: A speculative synthesis. Review. J Cereb Blood Flow Metabol 1:155–185
12. Todd MM, Chadwick HS, Shapiro HM, Dunlop BJ, Marshall LF, Dueck R (1982) The neurologic effects of thiopental therapy following experimental cardiac arrest in cats. Anesthesiology 57:76–86

Heidelberg, May 1983

Klaus Wiedemann,
Siegfried Hoyer

Contents

X

List of Authors*

Ashton, D. *100*
Astrup, J. *31*
Auer, L. M. *124*
van Belle, H. *76*
Borgers, M. *12*
Bosma, H.-J. *67*
Bruce, D. A. *158*
Clinke, G. *100*
Dearden, N. M. *112*
Gaab, M. R. *134*
Guggenberger, H. *38*
Heiss, W.-D. *25*
Heller, V. *134*
Hempelmann, G. *116*
Hermans, C. *76, 100*
Heuser, D. *38*
Hossmann, K. A. *67*
Hoyer, S. *81*
Jagenau, A. *76*
Kalimo, H. *1*
Kessler, P. D. *55*
Kling, D. *116*
Krier, C. *81*

Lin, M. R. *55*
van Loon, J. *76*
McDowall, D. G. *112*
Mies, G. *67*
Miller, J. D. *95*
Mullie, A. *76*
Nemmer, J. P. *45*
Nemoto, E. M. *45, 55*
Olsson, Y. *1*
Paljärvi, L. *1*
Paschen, W. *67*
Poch, B. *134*
van Reempts, J. *12*
Rosner, G. *25*
Russ, W. *116*
Shiu, G. K. *45*
Siesjö, B. K. *1*
Sold, M. *134*
Symon, L. *88*
Vandevelde, K. *76*
Wauquier, A. *76, 100*
Wiedemann, K. *146*
Winter, P. M. *45*

---

* For author's addresses refer to the according contribution heading

Structural Aspects of Energy Failure States in the Brain

H. Kalimo[1], L. Paljärvi[1], Y. Olsson[2], and B. K. Siesjö[3]

1 Department of Pathology, University of Turku, 20520 Turku 52, Finland
2 Neuropathological Laboratory, Institute of Pathology, University of Uppsala, 75122 Uppsala, Sweden
3 Laboratory of Experimental Brain Research, University of Lund, 22185 Lund, Sweden

General aspects

The energy produced in the brain is used for three general purposes. First, it is used to support transmission of electrical impulses, including events involved in synaptic activity. This work encompasses both the constant repumping of ions that conduct the currents during nervous activity and the synthesis of neurotransmitters and neuromodulators. This obviously requires a great proportion of the energy produced, but transmission is evidently not a vital function for the nerve cell itself. Thus, with marginal degrees of energy failure (e.g. due to hypoxia, ischemia or hypoglycemia), the neuron may become electrically silent and yet recover completely, if an adequate energy state is restored (see Astrup in this book). The structural detection of cellular injury necessitates visible changes, such as altered form, relationships or stainability of cells. It may be assumed with great confidence (though direct evidence is lacking) that the neurons passing the threshold of "transmission failure" do not show any structural alterations.

Second, even if signals are not transmitted, the maintenance of a normal intra- and extracellular mileu requires active, energy consuming pumping of ions (and other metabolites) across the cell membranes. The failure of these pumping systems, which signifies a more severe degree of energy failure, jeopardizes the life of the neuron (for further details see Siesjö (32), Astrup (3), Astrup in this book) as it would threaten any other cell (12). Possibly the adverse effects of such a failure are to a large extent due to influx of Ca^{2+} into cells, or its release from intracellular sequestration sites (32). After this second threshold is passed ("membrane failure", see Astrup in this book), the volumes of different cellular compartments may change, but they do not necessarily have to do so. Thus, swelling or shrinkage of cells and/or their organelles may reveal that the brain cells have incurred an injury, which may become irreversible, unless the energy supply is quickly restored. Furthermore, at this stage the texture of cells may change from the normal, revealing the ill effects (see below).

Third, the metabolic functions necessary to maintain the structural integrity of the neurons also consume energy. Among such functions, resynthesis of membrane constituents during their normal turnover, as well as intracellular transport may be critical targets of the injury. These functions are certainly vital for the neuron even though their impairment might not be expressed immediately, and such an injury may not become structurally apparent until much later, perhaps requiring days to manifest itself (8,19,26).

Dark type of ischemic nerve cell injury

The classical example of nerve cell injury that becomes structurally striking by marked shrinkage and swelling of the cells is the dark type of ischemic nerve cell injury ("ischemic cell change = ICC"). This injury begins with microvacuolation of the cytoplasm, which in electron microscope can be identified as ballooning of mitochondria. This is followed by progressive condensation of the neurons and swelling of perineuronal astrocytic processes (Fig. 1 and 4) resulting in dark triangular, shrunken neurons surrounded by clear vacuoles. Following the appearance of the so-called incrustations, the injured neurons will finally break down and disappear (4,5), i.e. the severe form of this type of injury is irreversible. The dark type of nerve cell injury is seen for example following incomplete and transient combined hypoxia and ischemia (Levine model, 5), in regional ischemia (middle cerebral artery clipping, 7) and after complete cerebral ischemia followed by recirculation (2,10).

Pale type of ischemic nerve cell injury

The widely used term "ischemic cell change" for the dark type of ischemic nerve cell injury is unfortunate, since in some other forms of ischemia a different, pale type of ischemic nerve cell injury is encountered. This type is characterized by increased electron lucency of the cells, clumping of nuclear chromatin and cell sap, as well as slight dilatation of RER cisternae and mitochondria (Fig. 2, 3, 5 and 6). It is a characteristic change following complete cerebral ischemia without recirculation (2,9,13,14), but it also has been described in a number of other models of ischemia (see below).

Pathogenesis of the acute ischemic cell injury

The presence of two different types of ischemic nerve cell injury, dark and pale, in different experimental models has prompted the pathogenetic proposal that flow of fluid irrigating the neurons injured by hypoxia is decisive in the development of the dark type of injury (6). In complete cerebral ischemia, the ion and water fluxes across the leaky membranes occur only locally as no circulation exists. Thus, no major shrinkage or swelling can take place (Fig. 7B), whereas, if the ischemia is incomplete or recirculation is allowed, the residual or restored flow of water and ions to the damaged areas will result in major volumetric changes of cells and their organelles (Fig. 7C).

3

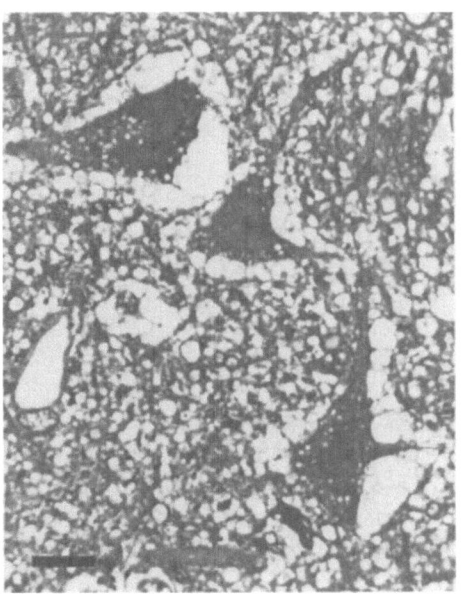 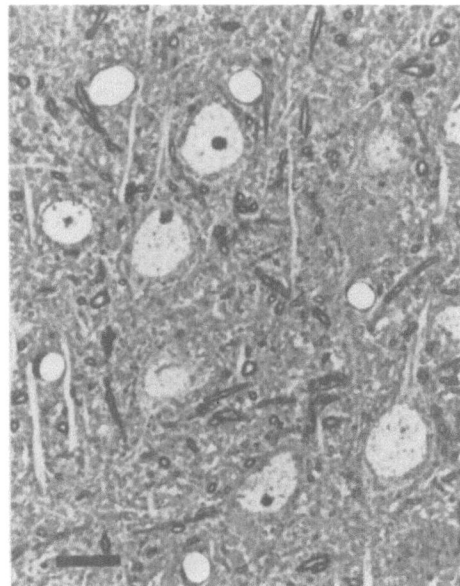

Fig. 1   Fig. 2

Fig. 1   In the dark type of ischemic injury, the nerve cell becomes triangular with marked condensation of the karyo- and cytoplasm, the latter containing micro-vacuoles. The injured neurons are surrounded by vacuoles and the neuropil appears markedly spongy due to the astrocytic edema. Sample from cerebral cortex of a rat exposed to four vessel occlusion ischemia (the model of Pulsinelli and Brierley, 25) for 30 min followed by recirculation for 90 min. Epon + toluidine blue, bar 10 µm

Fig. 2   Pale type of ischemic nerve cell injury in the cerebral cortex of a fasted rat exposed to 30 min of severe incomplete ischemia (CBF below 5% of normal) followed by 5 min of recirculation (16). The lactate level rises to about 15 µmol g^{-1} during the ischemic insult. Note the discrete changes with slightly coarsened nuclear chromatin in both neurons and astrocytes (A). No definite edema is detectable. Epon + toluidine blue, bar 10 µm

However, two recent findings point out that the pathogenesis of the dark nerve cell injury is more complex. In severe incomplete ischemia (cerebral blood flow below 5% of normal for 30 min) followed by recirculation, the previous hypothesis would predict the dark type of injury to appear. Yet, the structural changes were mainly of the pale type (Fig. 5 and 6) (16), even in the experimental group with excessive lactic acidosis, in which the neurons had obviously suffered an irreversible damage as indicated by the neurophysiological and metabolic results (29). A tentative explanation for the absence of the dark type of nerve cell injury is the following (for further details see Siesjö, 32 and Kalimo, 16):

In those models where the dark type of injury develops, the neuronal condensation is most often associated with massive swelling of the perineuronal astrocytic processes, which

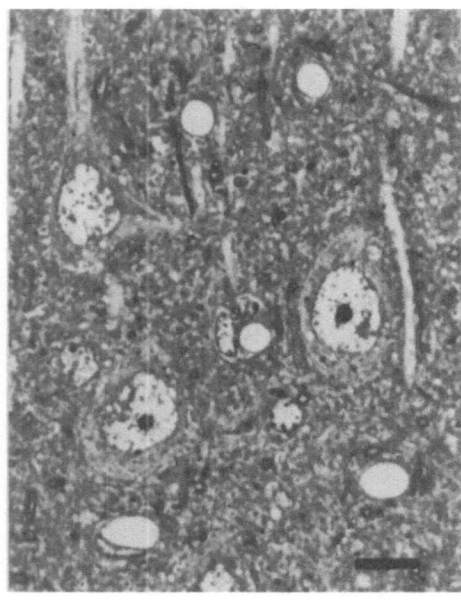

Fig. 3 Another pale type of injury in a rat exposed to similar ischemic insult as in Fig. 2, but with glucose pretreatment; the lactate rose to about 35 μmol g^{-1} during the insult. The clumping of the nuclear chromatin is more pronounced, but otherwise the changes are similar as in Fig. 2. Epon + toluidine blue, bar 10 μm.

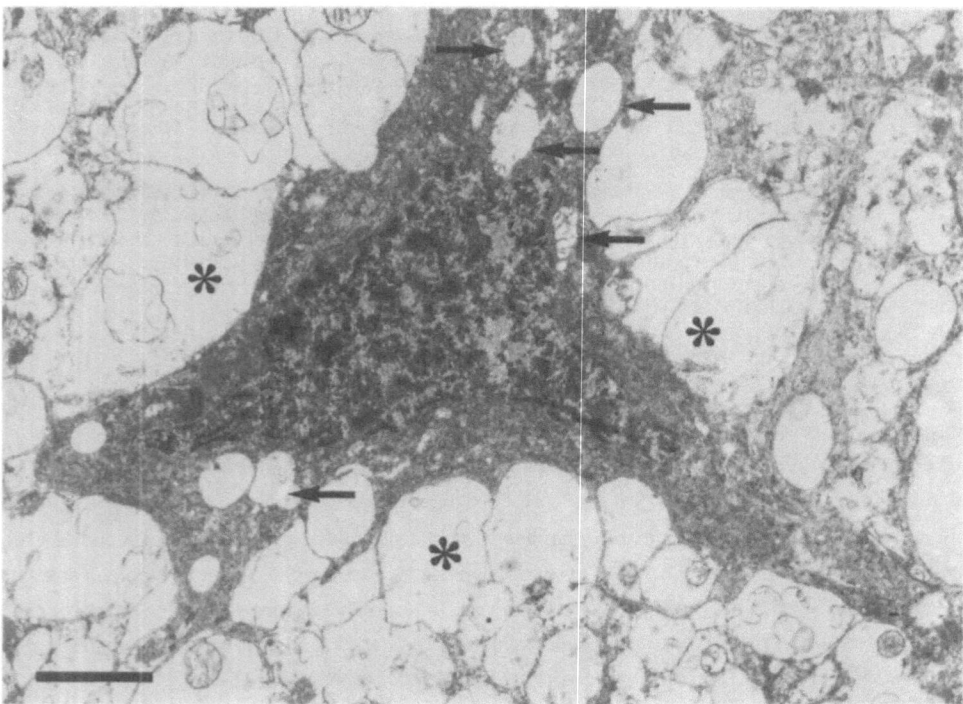

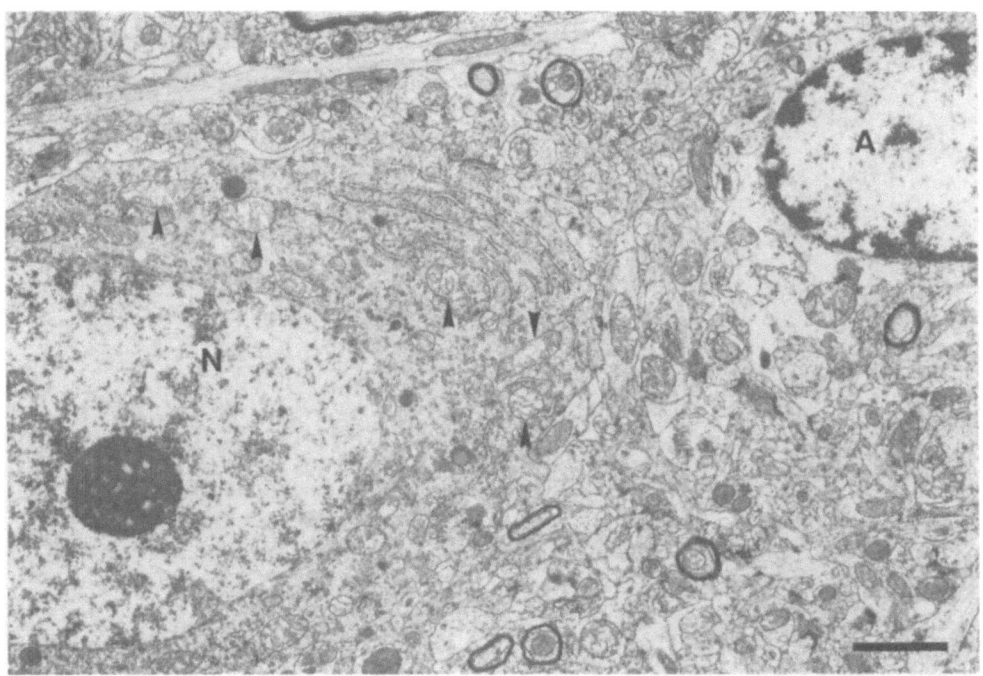

Fig. 5   Electron micrograph of the pale type of ischemic nerve cell injury from a rat
              exposed to ischemia as in Fig. 2. The nuclear chromatin in the neuron (N) and
              astrocytes (A) is coarsened. Neuronal mitochondria (arrows) are slightly swollen
              but not ballooned. Astrocytic cytoplasm appears somewhat edematous. (Re-
              produced with permission from Kalimo et al., 16). Bar 2 µm

Fig. 4   Dark type of ischemic nerve cell injury in EM from a rat exposed to similar
              ischemia as in Fig. 1. The condensation of the neuron is evident. The cytoplasmic
              microvacuoles can be identified as ballooned mitochondria (arrows). The peri-
              neuronal astrocytic processes are markedly swollen (asterisks). (Reproduced with
              permission from Kalimo et al., 17). Bar 2 µm

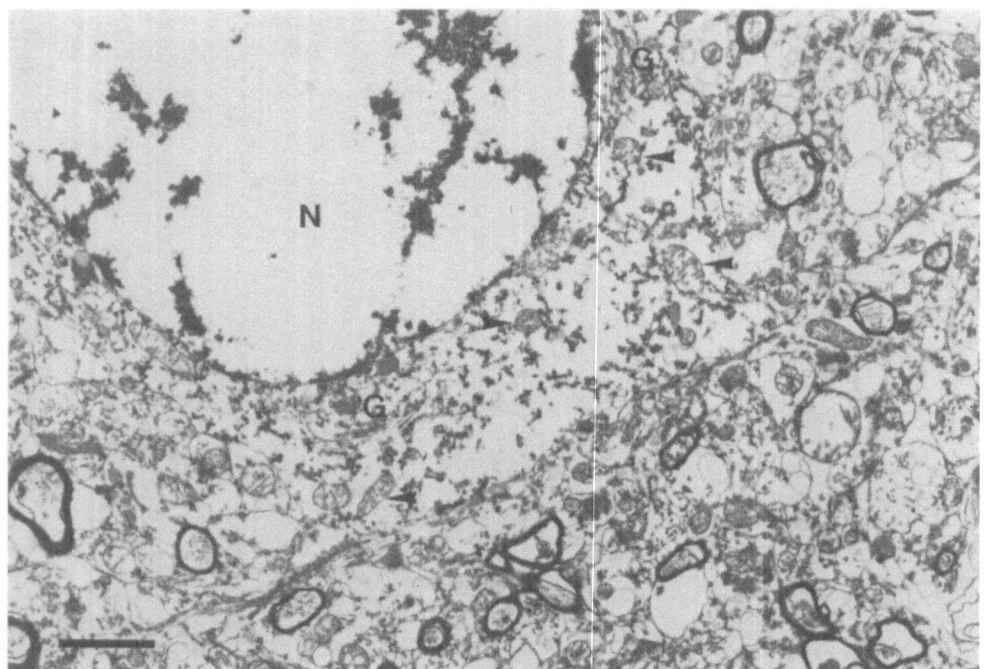

Fig. 6   Pale type of injury in the high lactic acidosis group as in Fig. 3. The nuclear chromatin (N) and cell sap of the neuron are severely clumped. The mitochondria (arrows) show only slight to moderate swelling. The surrounding neuropil is diffusely edematous. (Reproduced with permission from Kalimo et al., 16). Bar 2 μm

suggests a shift of osmotic equivalents and water from neurons to glial cells. It is known that if the extracellular K^+ concentration rises above $10 \mu mol.ml^{-1}$, the astrocytes are stimulated to actively take up Na^+ and Cl^- ions accompanied by water with consequent swelling (18). It may be assumed that in incomplete or transient cerebral ischemia (with residual or restored circulation, respectively) the more resistant astrocytes could outlive the more vulnerable neurons. In support, the swollen astrocyte processes around the dark type of injured neurons often contain well preserved mitochondria. Thus, these surviving astrocytes could be activated by the K^+ ions leaking from the more severely injured neurons which have passed the threshold of membrane failure, with resultant swelling of astrocytes and shrinkage of neurons (Fig. 7C).

In the above mentioned study on severe incomplete cerebral ischemia (16), the irreversibly injured group developed excessive tissue lactic acidosis during the ischemic insult due to pretreatment with glucose. The extreme lactacidosis may also have caused an irreversible injury to the astrocytes, and thus neither the active astrocytic swelling nor the consequent neuronal condensation could occur (Fig. 7D). This hypothesis thus makes a presumption that neurons can be selectively destroyed with astrocytes and vascular elements surviving, a view which was recently expressed by Marcoux et al. (20) as well as by Petito and Babiak (23), and which was explicitly stated by Plum (24) by saying that an infarction does not develop unless the astrocytes are also necrotized.

The other recent study demonstrating the complexity of the dark type of ischemic nerve cell injury is that by Jenkins and Becker (11). In their model of complete cerebral ischemia followed by recirculation, the dark type of ischemic nerve cell injury developed only if the brain was recirculated with whole blood. Reperfusion with saline or a solution containing oxygenated fluorochemicals did not give rise to dark injury, indicating that water, electrolytes or oxygen are not sufficient to elicit such structural changes. Since the dark type of injury is, at least in its severe form, irreversible, further studies are warranted to clarify the pathogenetic mechanisms involved.

Dark, condensed neurons (usually without the mitochondrial swelling) are also seen in other injurious states, which are commonly associated with tissue edema, e.g. in hypoglycemia (1) and status epilepticus (33). Thus, condensation of neurons seems to be closely related to astrocytic swelling, which suggests an interdependence of these two phenomena as well as a non-specific character of the neuronal shrinkage (but perhaps not of the mitochondrial swelling).

The role of lactic acidosis

The studies of Myers (21), Siemkowicz and Hansen (30), and Rehncrona et al. (28,29) have clearly shown in experimental animals that the amount of lactate accumulating in the brain of experimental animals crucially influences the severity of the ischemic tissue damage. Excessive lactate accumulation is structurally manifest in the severe pale type of cortical cell injury as more marked clumping of the nuclear chromatin and cell sap (cf. Fig. 5 and 6) in the high than in the low lactic acidosis group (16). The deleterious effect of the lactic acidosis has also been demonstrated clinically in the worse prognosis of stroke in both diabetic and nondiabetic patients with hyperglycemia (27). Siemkowicz et al. (31) propose that hyperglycemia may be detrimental even postischemically by inhibiting oxygen consumption and thus impairing recovery but their findings are at variance with those reported by Rehncrona et al. (28,29). The mechanism by which the excessive lactic acidosis causes the cell damage is still unclear, but it seems highly probable that it is related to the reduction of the intracellular pH. In the severe incomplete ischemia with excessive lactacidosis, the pH must fall well below 6. In the presence of energy failure, this may well cause irreversible denaturation of cell constituents, structurally visible as clumping of nuclear chromatin and cell sap. In contrast, such clumping is not seen in severe hypoglycemia (1,15) in which energy failure also occurs but no acidosis develops; rather, the intracellular pH rises (22).

8

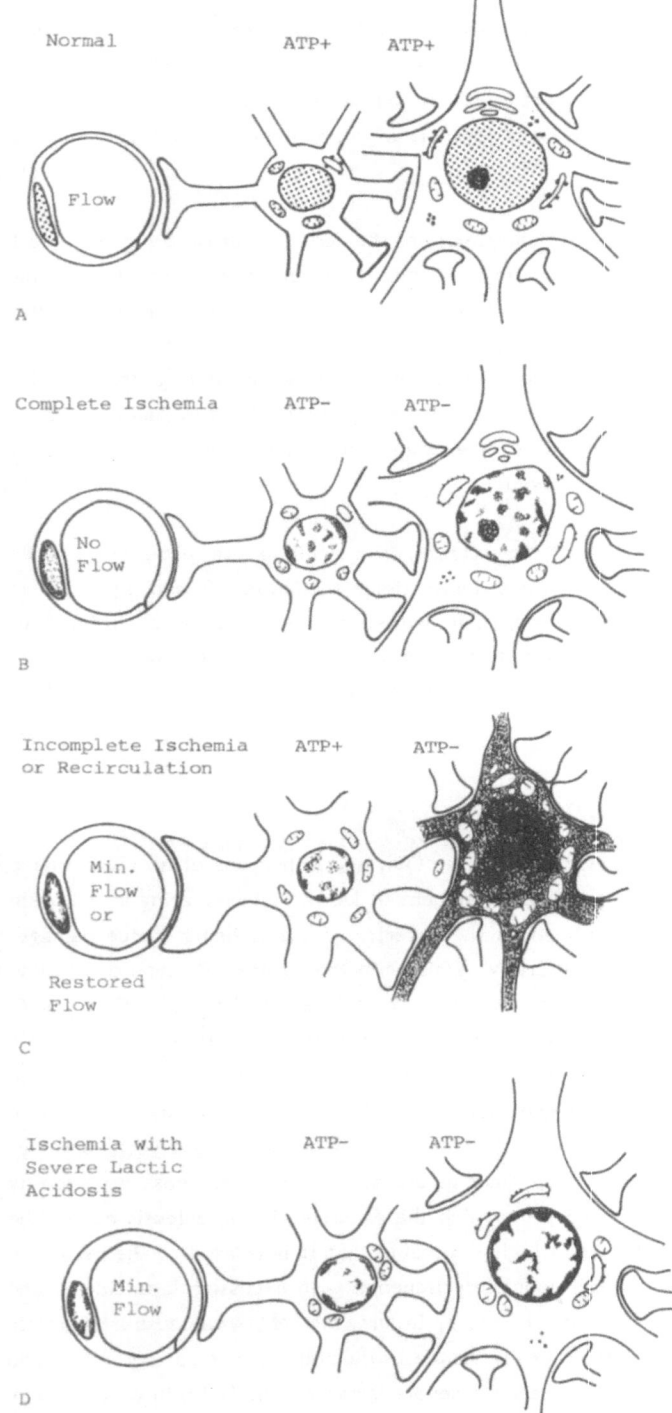

Fig. 7A   Schematic presentation of the structural changes in different forms of cerebral ischemia with pathogenic hypotheses. A. Normal situation

Fig. 7B   In complete cerebral ischemia, the energy failure renders the cell membranes leaky, but the ion and water gradients are leveled only locally, since no additional fluid enters the tissue. Thus, the cells or their organelles swell only slightly, while the extracellular space becomes narrowed

Fig. 7C   In incomplete ischemia, the astrocytes may survive, or after complete ischemia with recirculation they may recover, whereas in both cases the neurons remain injured. This result in marked astrocytic swelling and consequent neuronal shrinkage

Fig. 7D   In severe incomplete ischemia with excessive lactic acidosis, it is suggested that, at the low pH, also astrocytes incur an irreversible injury. Thus, they cannot actively swell and no major volume changes will take place

Conclusions

In conclusion, structural changes are not to be expected with such moderate energy failure that only abolishes neuronal transmission. Thus, structurally visible injury seems to require energy failure of a degree that affects ion and water compartmentation. Two main types of acute structural changes are seen in ischemically injured neurons. In the dark type of injury, shrinkage of neurons with ballooned mitochondria is accompanied by marked astrocytic swelling, whereas in the pale type of injury no major volume changes occur. It is proposed that development of the dark injury requires surviving astrocytes which swell actively and neurons consequently become condensed. Similar changes are seen in some other conditions with tissue edema (e.g. hypoglycemia and epilepsy), and thus neuronal shrinkage and astrocytic swelling are not specific for ischemia alone. In the pale type of injury, the astrocytes are also injured in such a way that they do not swell markedly. Excessive tissue lactic acidosis during the ischemic insult aggravates the ischemic damage and thus it should obviously be avoided to improve postischemic recovery.

References

1.   Agardh C-D, Kalimo H, Olsson Y, Siesjö BK (1980) Hypoglycemic brain injury. I. Metabolic and light microscopic findings in rat cerebral cortex during profound insulin-induced hypoglycemia and in the recovery period following glucose administration. Acta Neuropathol 50:31-41

2.   Arsenio-Nunes ML, Hossmann KA, Farkas-Bargerton E (1973) Ultrastructural and histochemical investigation of the cerebral cortex of cat during and after complete ischemia. Acta Neuropathol 26:329-344

3.   Astrup J (1982) Energy-requiring cell functions in the ischemic brain. Their critical supply and possible inhibition in protective therapy. J Neurosurg 56:482-497

4.   Brierley JB (1976) Cerebral hypoxia. In: Blackwood W, Corsellis JAN. Eds. Greenfield's Neuropathology. Edward Arnold, pp 43-85

5.   Brown AW, Brierley JB (1972) Anoxic-ischaemic cell change in rat brain. Light microscopic and fine structural observations. J Neurol Sci 16:59-84

6.   Garcia JH, Kamijyo Y, Kalimo H, Tanaka J, Viloria JE, Trump BF (1975) Cerebral ischemia: The early structural changes and correlation of these with known metabolic and dynamic abnormalities. In: Whisnant JP, Sandok B. Eds. Cerebral vascular diseases. Grune & Stratton, pp 313-323

7.   Garcia JH, Kalimo H, Kamijyo Y, Trump BF (1977) Cellular events during partial cerebral ischemia. I. Electron microscopy of feline cerebral cortex after middle-cerebral artery occlusion. Virchows Arch Cell Pathol 25:191-206

8.   Ito U, Spatz M, Walker JT jr, Klatzo I (1975) Experimental cerebral ischemia in Mongolian gerbils. Acta Neuropathol 32:209-223

9.   Jenkins LW, Povlishock JT, Becker DP, Miller JD, Sullivan HG (1979) Complete cerebral ischemia. An ultrastructural study. Acta Neuropathol 48:113-125

10.   Jenkins LW, Povlishock JT, Lewelt W, Miller JD, Becker DP (1981) The role of postischemic recirculation in the development of ischemic neuronal injury following complete cerebral ischemia. Acta Neuropathol 55:205-220

11.   Jenkins LW, Becker DP (1982) A quantitative analysis of glial swelling and ischemic neuronal injury following complete cerebral ischemia. In: The proceedings of the 5th International Symposium on Brain Edema, Groningen, The Netherlands, in press

12.   Jennings RB, Ganote CE, Reimer KA (1975) Ischemic tissue injury. Am J Pathol 81:179-198

13.   Kalimo H, Garcia JH, Kamijyo Y, Tanaka J, Trump BF (1977) The ultrastructure of "brain death". II. Electron microscopy of feline cortex after complete ischemia. Virchows Archiv Cell Pathol 25:207-220

14.   Kalimo H, Paljärvi L, Vapalahti M (1979) The early ultrastructural alterations in the rabbit cerebral and cerebellar cortex after compression ischaemia. Neuropathol Appl Neurobiol 5:211-223

15.   Kalimo H, Agardh C-D, Olsson Y, Siesjö BK (1980) Hypoglycemic brain injury. II. Electron microscopic findings in rat cerebral cortical neurons during profound insulin-induced hypoglycemia and in the recovery period following glucose administration. Acta Neuropathol 50:43-52

16.   Kalimo H, Rehncrona S, Söderfeldt B, Olsson Y, Siesjö BK (1981) Brain lactic acidosis and ischemic cell damage: 2. Histopathology. J Cereb Blood Flow Metabol 1:313-327

17.   Kalimo H, Paljärvi L, Olsson Y (1982) Morphological and biochemical features of brain hypoxia-ischemia. In: Wauquier A. Ed. Protection of tissues against hypoxia. Elsevier, in press

18.   Kimelberg HK, Bourke RS, Stieg PE, Barron KD, Hirata H, Pelton EW, Nelson LR (1982) Swelling of astroglia after injury to the central nervous system: Mechanisms and consequences: In: Grossman RG, Gildenberg PL. Eds. Head injury: Basic and clinical aspects. Raven, New York, pp 31-44

19.   Kirino T (1982) Delayed neuronal death in the gerbil hippocampus following ischemia. Brain Res 239:57-69

20. Marcoux FW, Morawetz RB, Crowell RM, DeGirolami U, Halsey JH (1982) Differential regional vulnerability in experimental focal cerebral ischemia. J Cereb Blood Flow Metabol 2:263

21. Myers RE (1977) Lactic acid accumulation as a cause cf brain edema and cerebral necrosis resulting from oxygen deprivation. In: Korobkin R, Guilleminault G. Eds. Advances in Perinatal Neurology. Spectrum, pp 85-114

22. Pelligrino D, Alquist LO, Siesjö BK (1981) Effects of insulin-induced hypoglycemia on intracellular pH and impedance in the cerebral cortex of the rat. Brain Res 221:129-147

23. Petito CK, Babiak T (1982) Early proliferative changes in astroctyes in postischemic noninfarcted rat brain. Ann Neurol 11:510-518

24. Plum F (1982) What causes infarction in ischemic brain? Wartenberg lecture 1982, and personal communication

25. Pulsinelli WA, Brierley JB (1979) A new model of bilateral hemispheric ischemia in the unanesthetized rat. Stroke 10:267-272

26. Pulsinelli WA, Brierley JB, Plum F (1982) Temporal profile of neuronal damage in a model of transient forebrain ischemia. Ann Neurol 11:491-498

27. Pulsinelli WA, Levy DE, Sigsbee B, Scherer P, Plum F (1982) Hyperglycemia with or without established diabetes mellitus worsens stroke outcome. Am J Med, in press

28. Rehncrona S, Rosén I, Siesjö BK (1980) Excessive cellular acidosis: An important mechanism of neuronal damage in the brain? Acta Physiol Scand 110:435-437

29. Rehncrona S, Rosén I, Siesjö BK (1981) Brain lactic acidosis and ischemic cell damage: 1. Biochemistry and neurophysiology. J Cereb Blood Flow Metabol 1:297-311

30. Siemkowicz E, Hansen AJ (1978) Clinical restitution following cerebral ischemia in hypo-, normo- and hyperglycemic rats. Acta Neurol Scand 58:1-8

31. Siemkowicz E, Hansen AJ, Gjedde A (1982) Hyperglycemic ischemia of rat brain: the effect of post-ischemic insulin on metabolic rate. Brain Res 243:386-390

32. Siesjö BK (1981) Cell damage in the brain: A speculative synthesis. J Cereb Blood Flow Metabol 1:155-185

33. Söderfeldt B, Kalimo H, Olsson Y, Siesjö BK (1981) Pathogenesis of brain lesions caused by experimental epilepsy. Light-and electron-microscopic changes in the rat cerebral cortex following bicuculline-induced status epilepticus. Acta Neuropathol 54:219-231

Morphological Aspects of Brain Protection in Experimentally Induced Hypoxia

J. van Reempts and M. Borgers

Laboratory of Cell Biology, Janssen Pharmaceutica, 2340 Beerse, Belgium

Introduction

The pathophysiology of brain hypoxia or ischemia has been the subject of many experimental studies because of its great importance in clinical fields such as critical care medicine, neurology and geriatrics.

Functional damage to neuronal and glial cells due to impairment of the cerebral circulation often results in important neurological deficits. These may be limited to inconspicious motor or sensory defects, but may also lead to severe dementia, seizures and even death. Only in the last decade has the conviction grown that brain injury after oxygen deprivation for periods longer than 4 min can be prevented. The belief that the brain can be protected and resuscitated not only results from a better understanding of brain pathophysiology but also from an improved technology and the development of pharmacological agents with brain protective properties. Although several classes of drugs are assumed to have such effects, the group of metabolic modifiers to which the barbiturates belong are the most widely investigated (1,33). Two drugs were recently added to the group of brain protective agents. Etomidate (Janssen Pharmaceutica) a short-acting nonbarbiturate hypnotic was introduced as an induction agent in anesthesia (14). The drug possesses anti-convulsant activity (3,34) and recently its antihypoxic action was reported (20,43). Flunarizine (Janssen Pharmaceutica), a calcium-entry blocker which has antivasoconstrictor properties (35) and which improves blood viscosity (12), has been reported recently to have antimigrainous (2) and antihypoxic activities (45,46).

The brain protective effects of both drugs were morphologically assessed in the Levine preparation, a model of unilateral hypoxic-ischemic cerebral damage (36,37). The aim of this paper is to compare morphological aspects of etomidate and flunarizine treatment in different hypoxia or ischemia models. Their effects on the cerebral circulation and their direct interaction with cell membranes might at least partly explain the observed beneficial action in alleviating brain injury during and after an hypoxic-ischemic insult.

Models for the histological study of hypoxic brain injury

The choice of an experimental model for the histological study of brain hypoxia and its pharmacological treatment must meet certain requirements. First, artificial alterations known as "dark neurons" and "hydropic cells" (9,10) have to be avoided since it is difficult to distinguish them from early experimentally induced changes. Adequate perfusion fixation is thus necessary and this can only be achieved when animals survive the hypoxic insult, so that the heart continues to beat until intracardiac perfusion starts. Moreover, the brains have to remain in the skull for an appropriate period of time to avoid mechanical damage to yet incompletely fixed areas (10). A second important requirement is that anesthetics should be avoided (28), since many agents used in anesthesia have brain protective properties. A third requirement is the use of small animals. This is merely a practical consideration, since ischemic damage to the brain is usually restricted to small areas, which in large brains are difficult to localize. On the other hand, the choice of a model should depend on the type of hypoxia or ischemia which one wants to study, or the therapeutic regimen which is persued. Global or focal, complete or incomplete and transient or permanent ischemia are different types of insult (30) and probably require different treatment schedules. For the above mentioned reasons, our studies were limited to the following small animal models which will be briefly described. The first model of hypoxic-ischemic brain damage, originally described by Levine (25) is based upon unilateral carotid artery ligation in rats followed by intermittent exposure of the conscious animals to pure nitrogen. In this way, animals survive and sufficient damage is produced to allow reliable histopathological evaluation at different survival periods and after different degrees of hypoxia (8,37,39). The damage found in this model is largely restricted to the ipsilateral hemisphere, particularly the 3rd, 5th and 6th layers of the parietal cortex and to a lesser extent to the pyramidal cell layer of the hippocampus. The second model is based upon cytotoxic hypoxia produced by intravenous injection in rats of 5 mg . kg^{-1} KCN. This produces an absence of reflexes and a flat EEG for 30 min. To keep these animals alive, artifical ventilation was initiated 4 min after cyanide injection, and continued until spontaneous breathing was restored (4). Damage was found almost exclusively in both hippocampi and was related to the duration of survival. The third model is one of transient ischemia in the gerbil, produced by bilateral carotid artery occlusion for 5 min followed by restoration of circulation and a survival period of 24 h (21). Damage was found bilaterally in the hippocampus and almost none in the cortex. The fourth model originally described by Pulsinelli (28) produces transient bilateral hemispheric ischemia in the unanesthetized rat and allowed survival periods of more than seven days. Ischemia is achieved by permanent occlusion of both vertebral arteries, followed the next day by a temporary occlusion of both carotid arteries. The lesions which develop are time-dependent and related to the duration of four-vessel occlusion. In the latter model, the hippocampus was the most vulnerable. When the morphological picture of cerebral damage in these models is compared, it turns out that, except for the Levine preparation, the hippocampus is the most vulnerable area of the brain. Moderate forms of damage, e.g. early chromatin clumping in astrocytes (Fig. 1A,B), sharply contrast with severe types of cell change. These can be classified in two distinct groups: coagulative necrosis, characterized by densification of the cellular cytoplasm and chromatin condensation or pyknosis of the nucleus (Fig. 1C) and edematous cell change, characterized by swelling of the cell body with associated disappearance of cytoplasmic organelles and nuclear chromatin clumping (Fig. 1D). Both types of change are often closely associated (23,24). Although it is difficult to classify extremely swollen, edematous cells, the electron microscopic observations point out

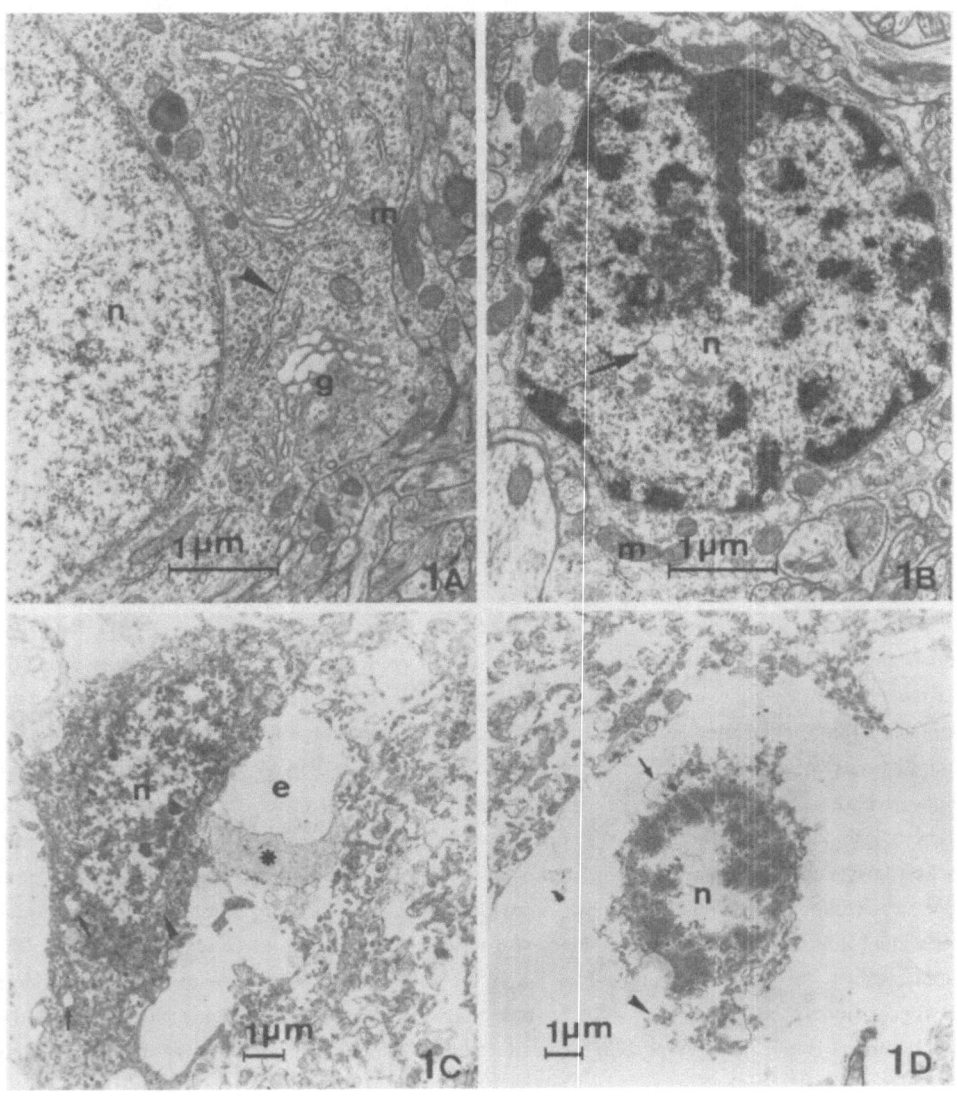

that both astrocytes and neurons are involved. Coagulative cell change on the other hand, seems to be restricted to neurons. It remains questionable, however, whether a particular morphological picture correlates with the type of ischemic-hypoxic insult.

Considerations on mechanisms of lesion formation

Interesting reviews on mechanisms of lesion formation have recently been published (17,32). In this paper, we will only discuss aspects related to the hypoxia models used for morphological study. When the balance between oxygen supply and oxygen demand is disturbed either by decreased blood flow, decreased arterial pO_2 or after cellular hyper-stimulation, or when the cells are deprived of necessary substrates such as glucose, a sequence of biochemical and physiological events begins, which results in loss of cellular function and finally cell death. Various pathological conditions may initiate cell death (e.g. toxins, ischemia, hyperstimulation), and because the final outcome is similar for all cells, a common final mechanism of destruction is most likely (15). The key position for calcium in this process has been proposed (7,15,26,32). Under normal conditions, calcium is an important regulator of many metabolic events and plays a key role in the maintenance of cell membrane integrity. During activation of the cells, calcium enters from the extra-cellular space and in order to regain the resting state, the cell calcium has to be removed, e.g. by energy-dependent membrane pumps. Energy failure leads to disturbance of the normal homeostasis for calcium and as a consequence of the strong inward electro-mechanical gradient which exists for this cation, the cell will become overloaded. This calcium accumulation in the cytosol can be demonstrated at the ultrastructural level with the oxalate-pyroantimonate technique originally described by Borgers et al. (6) and slightly modified for the retina and the brain (36). An example is given in Fig. 2 which shows calcium accumulation in a swollen cellular process in the hypoxic brain of a "Levine" rat. To verify

◄ Fig. 1   Ultrastructural pictures of cerebral cortical cells from hypoxic-ischemic rats, subjected to unilateral carotid artery ligation and exposed to pure nitrogen for nine periods of 1 min. A: Pyramidal neuron from an unaltered part of the third cortical layer. The nucleus (n) as well as the surrounding cytoplasm are well preserved. The Golgi cisternae (g) appear slightly dilated, while mitochondria (m) and endoplasmic reticulum (arrowhead) are completely normal. B: Protoplasmic astrocyte from the same are as in A in which signs of early damage can be seen. Condensation of karyoplasm is no longer restricted to the periphery of the nucleus (n) and the euchromatin shows early signs of vacuolation (arrow). Mitochondria (m) appear intact. C: Coagulative necrosis of a neuron from a severely damaged part of the cerebral cortex. The nucleus (n) shows a severe chromatin clumping and is surrounded by a dense shrunken cytoplasm which contains several vacuoles (arrows) and dilated mitochondria (arrowhead). The ischemic cell is surrounded by largely swollen astrocytic (?) cell processes (asterisk) and areas of severe extracellular edema (ed). D: Edematous cell change in a cortical cell from the same animal as in C. The nucleus (n) shows heavily condensed chromatin and is surrounded by a completely edematous cytoplasm in which only some vacuoles (arrow) or membranous remnants (arrowhead) can be recognized

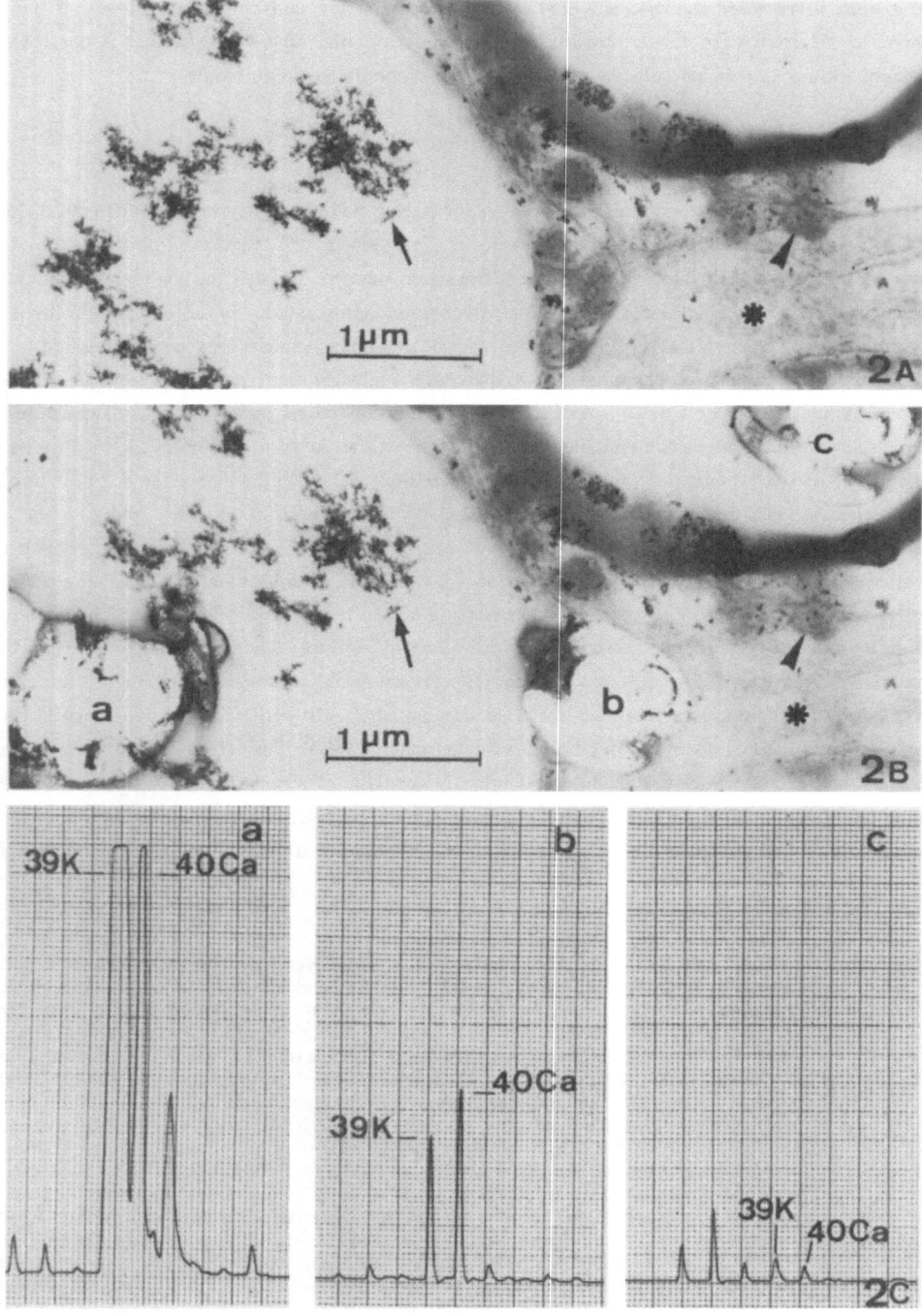

◄ Fig. 2   Ultrastructural detail on a 0.25 μm section of a severely damaged part of the cerebral cortex from a hypoxic-ischemic rat. Intracellular calcium overload was cytochemically demonstrated by means of a combined oxalate-pyroantimonate technique. Laser microprobe mass analysis was used to verify the nature of the detected precipitate. A: Accumulations of black precipitate can be seen in edematous cellular processes (arrow). Mitochondria (arrowhead) also show numerous electron-dense deposits, while the cellular cytoplasm (asterisk) is almost devoid of such precipitates. B: Appearance of the same section after evaporating small areas (a,b,c) with a high energy laser pulse. C: Mass spectra, corresponding to the three perforated areas, show elevated calcium peaks in a) and b), while calcium is almost absent in c). These values clearly correlate with the amount of cytochemically demonstrated precipitate which can be seen on the upper micrograph

the nature of the precipitate, a spectrum analysis was made by the LAMMA (Laser Microprobe Mass Analyzer) technique (41). This technique allows the determination of the exact composition of the precipitates in several subcellular compartments and has the advantage that the antimony signal does not interfere with the calcium signal. Moreover, the technique has a high sensitivity of about 10^{-20} g and a spacial resolution of about 1 μm. Calcium shifts during ischemia have also been measured with ion sensitive electrodes. These measurements have shown that ischemia induced a massive potassium efflux from brain cells and an associated decrease in extracellular calcium concentration (18,19). A close relationship exists between these ion changes and the rate of cerebral blood flow. At a flow of approximately 10 ml.100 g^{-1}.min^{-1}, there is a threshold below which ion homeostasis is disturbed (19). This value is very near the one which is related to a peak in edema formation. Indeed, it has been shown that ischemic edema is produced only when blood flow is below 20 ml.100 g^{-1}.min^{-1}, that it reaches a peak at 7 ml.100 g^{-1}.min^{-1} and that no edema is formed in areas of no flow (11). A first important aspect playing a role in the process of lesion formation is thus the rate of cerebral blood flow. Impaired circulation (ischemia) leads to insufficient delivery of energy substrates, although their concentration in the blood remains unaltered. A distinction should be made between different flow rates (i.e. low flow or no flow; reflow or no-reflow), each being responsible for a different degree of damage. Low flow is bad and progressively leads to edema formation, which in turn can perpetuate secondary ischemia by narrowing the vessel lumen. No flow, on the other hand, may be less harmful for a certain period of time, but the situation becomes especially harmful when blood flow is restored. In such a case, a short period of reactive hyperemia is followed by a delayed hypoperfusion period (17,32). During this recirculation period, brain protein synthesis is disrupted (27). The period of maximal hyperemia corresponds with a maximal edema formation (5). Here also secondary ischemia can be expected as a result from catecholamine-induced vasoconstriction and narrowing of the vascular lumen due to edema formation. An extreme form of secondary ischemia might then be complete occlusion of vessels in certain brain areas, which is thought to be responsible for the "no-reflow phenomenon" (16). Our own observations on regional microperfusion patterns during the early post-hypoxic recovery phase in Levine preparations (38) are in favor of the existence of delayed hypoperfusion or of a no-reflow phenomenon (Fig. 3). Although in the Levine preparation this effect is not related to transient ischemia followed by reperfusion, catecholamine liberation and formation of edema might be responsible for its occurrence.

A second important aspect in the mechanisms of lesion formation resides in the processes directly leading to energy deficit. In these situations, blood flow is normal but substrate provision is lowered (hypoxemia, hypoglycemia), substrate consumption is increased (status epilepticus) or cellular metabolism is impaired by acute toxicity of certain agents (cytotoxic hypoxia). Nevertheless, the maintenance of sufficient blood flow remains important, e.g. for the necessary removal of waste products. As already mentioned, the above described situations share one common mechanism of final cell death: a disturbed ion homeostasis, especially for calcium. Siesjö (32) proposes the following sequence of events in a condition of energy failure: efflux of K^+ and influx of Na^+ and Ca^{++}, uptake of K^+ and Cl^- by glial cells and consequent formation of astrocytic edema (22). The increased amount of calcium which penetrates the cytosol, initiates a series of additional deleterious processes, e.g. activation of calcium dependent phospholipases, which leads to destruction of membrane phospholipids and in turn results in free fatty acid accumulation. This probably is the reason why a common morphologic picture is encountered in conditions of ischemia, hypoglycemia or status epilepticus (32), namely the occurrence of coagulative neuronal necrosis associated with perineuronal astrocytic swelling.

The above described aspects of lesion formation are certainly not complete and although speculative, they are widely supported in recent literature. They may share a common mechanism finally leading to loss of function and cellular necrosis but, nevertheless, may require a different therapeutic approach.

Brain protective effects of etomidate and flunarizine

Prevention of brain injury can be achieved through prophylactic as well as curative treatment. Protection of the brain by these interventions can be afforded through different modes of action: by lowering metabolism, by providing adequate cerebral blood flow and by preserving ion homeostasis. To study the effects of anti-anoxic drugs, one has the choice between physiological, biochemical or cytochemical techniques, which may provide information on early events. Morphological techniques are well-suited to study relatively late events, such as irreversible cellular damage.

The effects of etomidate were investigated morphologically. The drug afforded a drastic inhibition or even complete prevention of coagulative necrosis as well as edematous cell change. This was seen in the Levine preparation (37) where after subcutaneous pretreatment with 10 mg.kg^{-1} etomidate, lesions were found in only three out of 16 rats, while at a lower

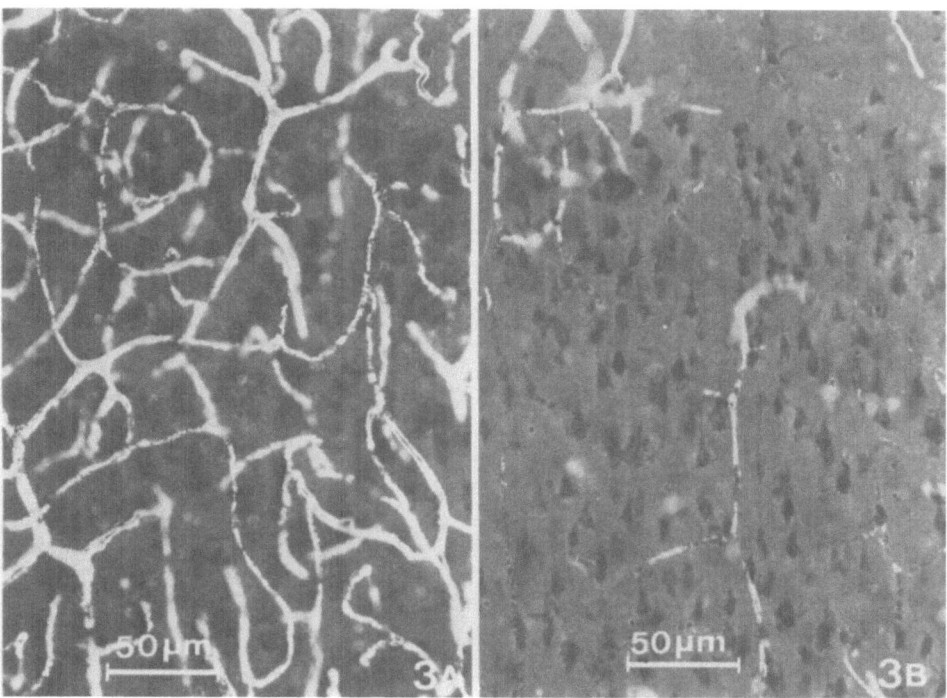

Fig. 3   Dark field picture of the microcirculation in the brain of a hypoxic-ischemic rat
subjected to unilateral carotid artery ligation followed by intermittent exposure
of the animal to nitrogen and sacrificed 30 min after the last hypoxic period. The
cerebral perfusion was visualized by filling the circulatory system with liquid
photographic emulsion which, after preparation of 100 μm vibratome sections,
was developed to metabolic silver (40). In comparison to a normal perfusion
pattern which is seen in the contralateral hemisphere (A), areas of delayed
hypoperfusion can easily be recognized in the ipsilateral hemisphere due to the
absence of light reflecting silver grains (B)

dose of 2.5 mg.kg^{-1} lesions were found in eight out of 16 rats. The protection was
independent of the hypnotic effect of the drug (37,44). In the untreated group, all rats (8 out
of 8) had lesions. These results are now confirmed by ultrastructural studies (unpublished)
which show that in etomidate-treated rats, subcellular structure is well preserved (Fig. 4A).
Similar findings were reported for the effects of etomidate against cyanide dysoxia (4). A
dose of 10 mg.kg^{-1} given subcutaneously at 30 min before intravenous injection of 5 mg.kg^{-1}
KCN prevented damage to the pyramidal cell layer of the hippocampus in all treated rats. In
the untreated series, microvacuolation was found in seven out of 18 and ischemic cell
change in four out of 18 rats. Again these results were supported by ultrastructural
observations. Recently, histological findings suggested brain protective effects after
curative treatment with etomidate at the end of a 5 minute period of bilateral carotid

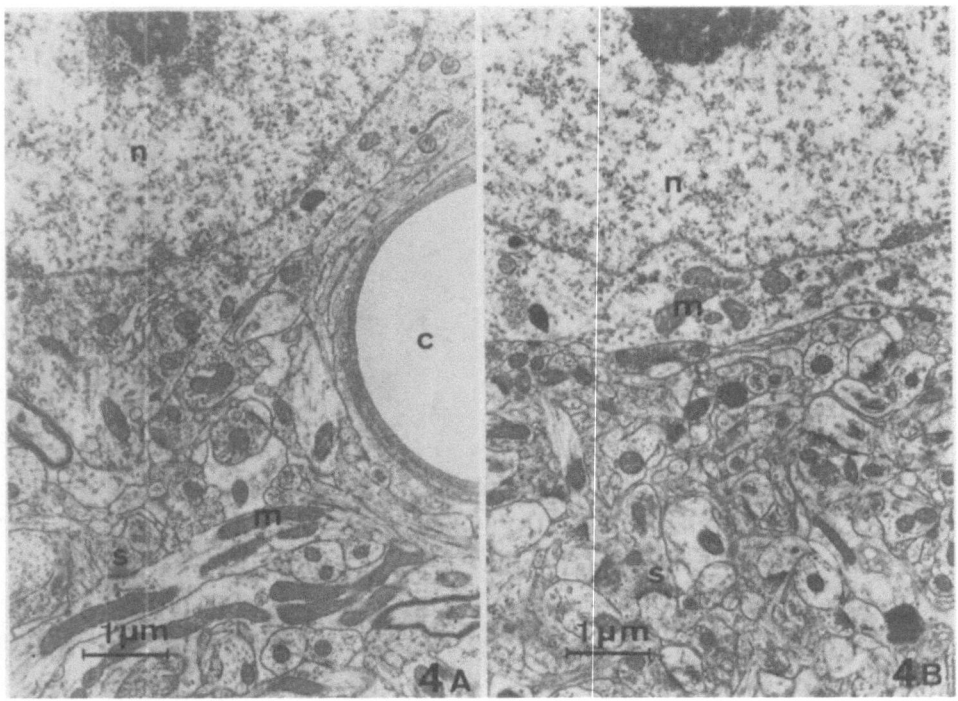

Fig. 4   Electron micrographs which show the anti-hypoxic effects of (A) 2.5 mg.kg^{-1} etomidate given subcutaneously at 1 h before nitrogen exposure and of (B) 10 mg.kg^{-1} flunarizine given orally at 5 h before hypoxia. The pictures are from the ipsilateral cortex of rats subjected to unilateral carotid artery ligation, followed by intermittent exposure to nitrogen for nine periods of 1 min. The ultrastructural appearance is comparable to that of normal rats. n = nucleus, m = mitochondria, s = synaptic ending, c = capillary

artery occlusion in the gerbil (21). At the present time, studies are in progress using the Pulsinelli model of four-vessel occlusion to further evaluate the effects of curative treatment with etomidate.

The antihypoxic properties of flunarizine were studied morphologically and cytochemically using the levine preparation. Light microscopic investigations showed a significant and dose-dependent reduction of both coagulative necrosis and edematous cell change after oral pretreatment with 10, 20 and 40 mg.kg^{-1} of flunarizine (39). At the ultrastructural level using the cytochemical oxalate-pyroantimonate technique for the detection of calcium (36), a normal distribution of this cation was found in treated animals. In contrast, untreated animals showed a great amount of intracellular calcium-antimonate precipitates (Fig. 2) (38). The subcellular preservation in treated rats was comparable to that of normal ones, confirming thereby the gross structural observations at the light microscopical level (Fig. 4B).

Aspects of the mechanism of action of etomidate and flunarizine

In the following, we will comment briefly on the above described morphological data and relate them to a possible mode of action of either drug.

For etomidate, it is clear that complete cellular destruction can be reduced or even prevented when the drug is given before the onset of hypoxia (37). Recent observations in the gerbil also suggest a possible curative effect of this drug against the deleterious consequences of the ischemic/postischemic recirculation insult (21). Not only neuronal cells survived the hypoxic or ischemic aggression, but also edematous cell swelling of astrocytes and certain dendritic processes was inhibited. A direct interference of etomidate with the regulatory systems of the cerebral circulation (i.e. by maintenance of adequate perfusion or shunting of blood to ischemic areas) may be envisaged. Etomidate which lowers the rise of intracranial pressure (31) might prevent reduced perfusion resulting from cerebral edema. Such a mode of action is supported by the finding that etomidate reduces edematous cell change and cell loss, both responsible for the spongy appearance of almost 50% of untreated hypoxic-ischemic rats (37). The importance of blood pressure recordings in Levine preparations may not be overlooked. It is possible that the sudden rise in systemic pressure which was observed during the early posthypoxic period (38) may be accompanied by local changes in cerebral blood flow, such as redistribution of flow to highly vulnerable areas. Interference of etomidate with the cellular metabolism could not be derived from the morphological observations. However, data are available which indicate a lowering effect on the cerebral metabolic rate of O_2 consumption (29). Due to the hypnotic effect of etomidate, the total O_2 consumption in the whole body might be lowered and thus might delay deleterious mechanisms (20). On the other hand, experimental data point out that the hypnotic effect does not seem to be necessary to obtain protection (37,44) and thus do not support this assumption. The anti-convulsant properties of etomidate (3,34) might favor a mechanism of energy preservation in hypoxic conditions. However, it must be emphasized that protection has also been obtained in etomidate-treated animals in which the anti-convulsive activities had worn off (37).

The protective effects of flunarizine, on the other hand, might be attributed to direct interaction with the cell membrane components, thereby preventing deleterious calcium overload of the cytosol. Prevention of calcium intoxication after hypoxic-ischemic aggression was shown cytochemically (38). Apart from the above, it is not excluded that flunarizine interferes with the circulatory system. Its anti-vasoconstrictor effects (35) and its improvement of microrheological behavior (12) might maintain a sufficient cerebral flow and in this way prevent edema formation and the resulting secondary ischemia. Finally, the protective effects of flunarizine against cerebral hypoxic damage might partly be the result of its anti-convulsant activities (42). Again there is no direct evidence from our data with regard to this since flunarizine, although protective in the Levine preparation, did not prevent the type of convulsions encountered during nitrogen exposure (39).

In view of the above mentioned findings, it is difficult to draw definite conclusions concerning the exact mechanisms by which flunarizine and etomidate afford protection to the brain.

References

1.   Aitkenhead AR (1981) Do barbiturates protect the brain? Br J Anaesth 53:1011-1014

2.   Amery W, Wauquier A, Van Nueten J, De Clerck F, Van Reempts J, Janssen PAJ (1981) The anti-migrainous pharmacology of flunarizine (R 14 950), a calcium antagonist. Drugs Experiment Clin Res 7:1-10

3.   Ashton D, Wauquier A (1979) Effects of some anti-epileptic, neuroleptic and gabaminergic drugs on convulsions induced in rats by injection of D,L-allylglycine. Pharmacol Biochem Behav 11:221-226

4.   Ashton D, Van Reempts J, Wauquier A (1981) Behavioral, electroencephalographic and histological study of the protective effect of etomidate against histotoxic anoxia produced by cyanide. Arch Int Pharmacodyn Thérap 254:196-213

5.   Avery S, Crockard A, Ross Russell R (1982) Reperfusion oedema in gerbils. In: Proceed 5th Int Symp Brain Edema. Groningen, 1982, in press

6.   Borgers M, De Brabander M, Van Reempts J, Awouters F, Jacob WA (1977) Intranuclear microtubules in lung mast cells of guinea pigs in anaphylactic shock. Lab Invest 37:1-8

7.   Borgers M, Thoné F, Van Reempts J, Verheyen F (1982) The role of calcium in cellular dysfunction. In: Proceed Int Symp Cerebral Resuscit. Detroit, 1982, in press

8.   Brown AW, Brierley JB (1968) The nature, distribution and earliest stages of anoxic-ischaemic nerve cell damage in the rat brain, as defined by the optical microscope. Br J Exp Path 49:87-106

9.   Brown AW (1977) Structural abnormalities in neurons. J Clin Path (Suppl 11) 30:155-169

10.   Cammermeyer J (1978) Is the solitary dark neuron a manifestation of postmortem trauma to the brain inadequately fixed by perfusion? Histochemistry 56:97-115

11.   Crockard A, Ianotti F, Hunstock AT, Smith RD, Harris RJ, Symon L (1980) Cerebral blood flow and edema following carotid occlusion in the gerbil. Stroke 11:494-498

12.   De Clerck F, David JL (1981) Pharmacological control of platelet and red blood cell function in the microcirculation. J Cardiovasc Pharmacol 3:1388-1412

13.   Dienel GA, Pulsinelli WA, Duffy TE (1980) Regional protein synthesis in rat brain following acute hemispheric ischemia. J Neurochem 35:1216-1226

14. Doenicke A, Kugler J, Penzel G, Laub M, Kalmar L, Killian J, Bezecny A (1973) Hirnfunktion und Toleranzbreite nach Etomidate, einem neuen barbituratfreien i.v. applizierbaren Hypnotikum. Anaesthesist 22:357-366

15. Farber JL (1981) The role of calcium in cell death. Life Sci 29:1289-1295

16. Fischer EG, Ames A III, Hedley-Whyte ET, O'Gorman S (1977) Reassessment of cerebral capillary changes in acute global ischemia and their relationship to the "No-reflow phenomenon". Stroke 8:36-39

17. Frost EAM (1981) Brain preservation. Anesthesia Analgesia 60:821-832

18. Hanssen AJ (1981) Extracellular ion concentrations in cerebral ischemia. In: Zeuthen T. Ed. The application of ion-sensitive microelectrodes. Elsevier, Amsterdam, pp 239-254

19. Harris RJ, Symon L, Branston NM, Bayhan M (1981) Changes in cellular calcium activity in cerebral ischemia. J Cereb Blood Flow Metabol 1:203-209

20. Hempelmann G, Lüben V, Klug N (1982) Möglichkeiten der Hirnprotektion unter besonderer Berücksichtigung von Etomidate (Hypnomidat^R). Notfallmedizin 8:83-95

21. Hermans CFM, Van Reempts J, Wauquier A (1982) Neurological outcome, EEG and brain histology following post-treatment with etomidate and thiopental after bilateral carotid occlusion and reperfusion in the gerbil. In preparation

22. Hertz L (1981) Features of astrocytic function apparently involved in the response of central nervous tissue to ischemia-hypoxia. J Cereb Blood Flow Metabol 1:143-153

23. Jenkins LW, Becker DP (1982) A quantitative analysis of perivascular glial swelling and ischemic neuronal injury. In: Gwan Go K. Ed. Proceed 5th Int Symp Brain Edema, Groningen. Plenum Press, in press

24. Kalimo H (1982) Biochemical and morphological features of brain hypoxia-ischemia. Proceed Int Symp Protection of Tissues Against Hypoxia, Beerse, 1982, in press

25. Levine S (1960) Anoxic-ischemic encephalopathy in rats. Amer J Pathol 36:1-17

26. Meldrum B, Griffiths T, Evans M (1982) Hypoxia and neuronal hyperexcitability. A clue to mechanisms of brain protection. Proceed Int Symp Protection of Tissues Against Hypoxia, Beerse, 1982, in press

27. Nemoto EM (1978) Pathogenesis of cerebral ischemia-anoxia. Crit Care Med 6:203-214

28. Pulsinelli WA, Brierley JB (1979) A new model of bilateral hemispheric ischemia in the unanesthetized rat. Stroke 10:267-272

29. Renou AM, Vernhiet J, Macrez P, Constant P, Billerey J, Khadaroo MY, Caille JM (1978) Cerebral blood flow and metabolism during etomidate anesthesia in man. Br J Anaest 50:1047-1051

30. Safar P (1978) Introduction: on the evolution of brain resuscitation. Crit Care Med 6:199-202

31. Schulte am Esch J, Pfeifer G, Thiening F (1978) Der Einfluß von Etomidate und Thiopental auf den gesteigerten intracraniellen Druck. Anaesthesist 27:71-75

32. Siesjö BK (1981) Cell damage in the brain: a speculative synthesis. J Cereb Blood Flow Metabol 1:155-185

33. Steen PA, Michenfelder JD (1980) Mechanisms of barbiturate protection. Anesthesiology 53:183-185

34. Van der Starre P (1980) Etomidat als schnellwirkende anti-epileptische Substanz. In: Opitz A, Degen R. Eds. Anästhesie bei zerebralen Krampfanfällen und Intensivtherapie des Status Epilepticus. Verlagsgesellschaft, Erlangen, pp 205-208

35. Van Nueten JM, Van Beek J, Janssen PAJ (1978) Effect of flunarizine on calcium-induced responses of peripheral vascular smooth muscle. Arch Int Pharmacodyn Thérap 232:42-52

36. Van Reempts J, Borgers M, Offner F (1982) Ultrastructural localisation of calcium in the rat retina with a combined oxalate-pyroantimonate technique. Histochem J 14:517-522

37. Van Reempts J, Borgers M, Van Eyndhoven J, Hermans C (1982) The protective effects of etomidate in hypoxic-ischemic brain damage in the rat. A morphologic assessment. Exp Neurol 76:181-195

38. Van Reempts J, Borgers M (1982) Morphological assessment of pharmacological brain protection. In: Proceed Int Symp on Protection of Tissues Against Hypoxia, Beerse, 1982, in press

39. Van Reempts J, Borgers M, van Dael L, Van Eyndhoven J, Van de Ven M (1982) Protection with flunarizine against hypoxic-ischemic damage of the rat cerebral cortex. A quantitative morphologic approach. Arch Int Pharmacodyn Thérap, in press

40. Van Reempts J, Haseldonckx M, Borgers M (1982) A simple technique for the microscopic study of microvascular geometry and tissue perfusion, allowing simultaneous histopathologic evaluation. Microvascular Research, in press

41. Van Reempts J, Borgers M, De Nollin S, Garrevoet T, Jacob W Ultrastructural calcium localization in the brain using a combined oxalate-pyroantimonate method. Specificity control by Laser Microprobe Mass Analysis (LAMMA), in preparation

42. Wauquier A, Ashton D, Melis W (1979) Behavioural analysis of amygdaloid kindling in Beagle dogs and the effects of clonazepam, diazepam, phenobarbital, diphenyl-hydantoin and flunarizine on seizure manifestation. Exp Neurol 64:579-586

43. Wauquier A, Ashton D, Clincke G, Niemegeers CJE, Janssen PAJ (1980) Etomidate: ein barbituratfreies Hypnotikum: antikonvulsive, antianoxische, und hirnprotektive Wirkung in Tierexperiment. In: Opitz A, Degen R. Eds. Anästhesie bei zerebralen Krampfanfällen und Intensivtherapie des Status Epilepticus. Verlagsgesellschaft, Erlangen, pp 183-203

44. Wauquier A, Ashton D, Clincke G, Van Reempts J (1981) Considerations on models and treatment of brain hypoxia. In: Van Hof MW, Mohn S. Eds. Developments in Neuroscience: Recovery from brain damage. Elsevier, Amsterdam, pp 95-114

45. Wauquier A, Ashton D, Clincke G, Van Reempts J (1982) Pharmacological protection against brain hypoxia: the efficacy of flunarizine, a calcium entry blocker. In: Clifford Rose F., Amery WK Eds. Hypoxia in the pathogenesis of migraine attacks. Pitman, London, in press

46. White BC, Gadzinsky DS, Hochner DJ, Krome C, Hochner T, White J, Trombley JH (1982) Correction of canine cerebral cortical blood flow and vascular resistance after cardiac arrest using flunarizine, a calcium antagonist. Ann Emerg Med 22:118-127

Survival of Cortical Neurons After Ischemia: Dependency on Severity and Duration

G. Rosner and W.-D. Heiss

Max-Planck-Institut für Neurologische Forschung, Ostmerheimer Strasse 200, 5000 Köln 91, FRG

Introduction

The function of the brain is closely related to the cerebral blood supply. If the cerebral blood flow falls below a value of 0.2 ml/g/min, EEG frequency decreases and the amplitude of the EEG becomes isoelectric at about 0.15 ml/g/min (12,15). Somatosensory evoked potentials are nearly unchanged until flow drops below 0.2 ml/g/min. The amplitudes then become smaller and smaller until at flow values of about 0.15 ml/g/min, the evoked potentials are totally abolished (2,16). In addition, Umbach et al. (16) showed that the presynaptic and postsynaptic components of the evoked potentials have different sensitivity: the postsynaptic component decreases first and disappears at about 0.15 ml/g/min. The presynaptic component can be recorded down to a flow of 0.12 ml/g/min. The spontaneous activity of cortical cells changes at a flow below 0.3 ml/g/min and ceases at 0.18 ml/g/min (4). Morawetz et al. (9) found additionally that the duration of ischemia is also a determinant for the survival of brain tissue at various levels of flow disturbance. They showed that persistent infarcts were only produced after occlusions of 120 min with a residual flow of 0.12 ml/g/min. The aim of the present study was to analyze in detail the effect of both severity and duration of ischemia on the spontaneous activity of cortical nerve cells.

Methods

Adult cats with a body weight of 1.5 to 3.0 kg were used. After tracheotomy, animals were mechanically ventilated with a gas mixture of oxygen (30%) and nitrous oxide (70%) and sufficient carbon dioxide to maintain $PaCO_2$ at 27 to 33 mmHg. For determining the cerebral blood flow, the hydrogen clearance method (1) was used. The head of the cat was fixed in a stereotaxic headholder. The left middle cerebral artery (mca) was approached transorbitally, and a hook, prepared from a cannula (8,14), was inserted above the proximal segment of the mca. After occluding the orbita, the mca could be reversibly occluded for variable periods of time by a microdriven angiographic guide wire. After craniotomy, a glass-coated platinium-iridium microelectrode was inserted into the sylvian or middle ectosylvian gyrus. A macroelectrode was inserted close to the microelectrode. With the microelectrode, pericellular blood flow and neuronal activity were recorded simultaneously (10). The macroelectrode was used for measuring the mean blood flow. Action potentials were stored on magnetic tape. Action potential frequency was simultaneously registered on a penrecorder after integration with a sampling interval of 10 s.

Results

Relationship between cerebral blood flow and neuronal activity

Under control conditions, blood flow ranged from 0.6 to 0.8 ml/g/min and activity of the neurons showed an irregular pattern. If the flow was reduced below a value of 0.3 ml/g/min, the spontaneous activity changed to a more regular or bursting pattern. In most cortical cells, the spontaneous activity disappeared below values of about 0.18 ml/g/min. To further analyze the importance of this flow, threshold activity of two cells was recorded with one microelectrode. The two cells were distinguishable by different spike amplitudes differentiated with a window discriminator. Neurons with large spikes (peak to peak 100 µV to several mV) stopped their activity at a flow of 0.18 ml/g/min, while that of neurons with lower spikes (peak to peak 50 -100 µV) were only slightly diminished. These cells with spikes of small amplitudes lost their activity at lower flow values. The lowest value at which spontaneous activity could be observed was 0.06 ml/g/min.

Duration of ischemia as a determinant for irreversible neuronal damage

Depending on the residual flow and the duration of ischemia, the neurons recovered during reperfusion. Results from a representative neuron are shown in Fig. 1. This neuron stopped firing at a flow value of 0.07 ml/g/min. After the end of the occlusion, the pericellular blood flow showed the typical hyperperfusion (1.8 ml/g/min). In this case, time for recovery of 10 min was relatively short as compared to the mean recovery time of 25 min (Table 1). No neuron was observed which tolerated a reduced flow in the range of 0.05 to 0.08 ml/g/min longer than 30 min. Some cells did not return after 30 min occlusion with a residual flow higher than 0.12 ml/g/min, which is generally tolerated for longer durations. However, it has to be considered that due to the movements of the brain, these neurons may have been distorted from the recording electrode.

Recovery time after graded ischemia

The time for recovery of spontaneous neuronal activity depended on both duration and severity of ischemia. This is shown in Table 1, where all our experiments are summarized in three classes according to magnitude of residual flow. Each class was further differentiated into three classes depending on the duration of occlusion.

After a strong ischemia of below 0.08 ml/g/min lasting 13 min, the mean recovery time was 21 min. The time of reperfusion was 12 min after a residual flow of 0.08 - 0.11 ml/g/min and

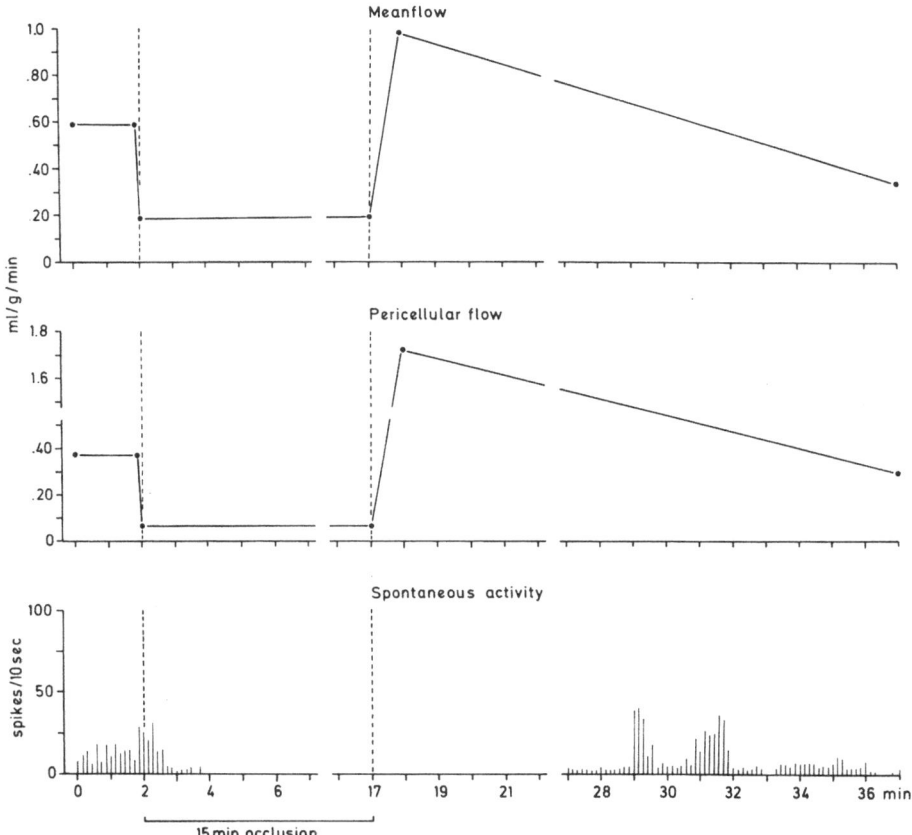

Fig. 1   Mean flow recorded with the macroelectrode, pericellular flow and spontaneous activity of a neuron before, during, and after 15 min ischemia (0.07 ml/g/min). The neuron's activity recorded with microelectrode is abolished during the ischemic period, but returns with reperfusion after 10 min

an occlusion time up to 30 min. If occlusion time was extended to up to 150 min at a flow of 0.12 - 0.2 ml/g/min, spontaneous activity of the neurons recovered after 26 min of reperfusion. This mean recovery time is comparable with that measured after a residual

flow below 0.09 ml/g/min and an occlusion time of 15 - 30 min. Thus the mean recovery time was similar for spontaneous activity of cerebral neurons after a short but severe and long-lasting but slight ischemia.

If residual flow during mca clamp exceeded 0.2 ml/g/min, the spontaneous firing of the cells either started again already during the clamp, or firing frequency was only diminished and upon reopening of the clamp immediately returned to the original value. Sometimes long periods of graded ischemia may be tolerated by cortical neurons: the longest duration of

Table 1   Reversibility of cellular spontaneous activity in relationship to
severity and duration of ischemia

	residual flow ml/g/min	duration min	n	mean time for recovery min
Class I	0.05 – 0.08	13	5	no recovery
			2	21
		15 – 30	7	25
		30 – 45	2	no recovery
Class II	0.09 – 0.11	13	1	11
		15 – 30	7	12
		30 – 90	3	no recovery
Class III	0.12 – 0.2	13	1	20
		15 – 30	1	no recovery
			8	13
		30 – 150	9	26
			5	no recovery

graded ischemia tested in our experiments was more than 2 h. In this case, a simultaneous recording from two cells was obtained. The neuron with the larger spike amplitudes stopped its activity at a pericellular blood flow of 0.17 ml/g/min and its activity returned after 32 min of reperfusion. The cell with the lower spike amplitudes was firing throughout the whole duration of occlusion though at a somewhat reduced frequency.

Discussion

Our results demonstrate that the survival of neurons depends on duration as well as severity of ischemia. Critical for the viability of a neuron are ischemic events with a residual flow below 0.05 ml/g/min for more than 20 min, or below 0.08 ml/g/min for more than 30 min, or below 0.14 ml/g/min for more than 45 min. From a borderline calculated from these values (5), it may be concluded that a residual flow of 0.18 ml/g/min is a critical value which can be tolerated for unlimited periods of time. The value of 0.18 ml/g/min observed previously (4) was in this study the median value for presistence of spontaneous activity of cortical neurons with a variability between 0.22 and 0.064 ml/g/min. This sequential dropout of the activity of nerve cells corresponds well to the gradual development of paralysis in the baboon when flow is decreased from 0.23 to 0.07 ml/g/min (7). In this varying threshold for cell firing, a selective sensitivity of the cells against ischemia is functionally demonstrated. This selective sensitivity could be directly observed when spontaneous activity of two different neurons was recorded simultaneously by one microelectrode. The selective ischemic vulnerability of cells (11,13) was also represented in the broad overlap of viable and devitalized cells with respect to severity and duration of ischemia. Selective vulnerability to ischemic or anoxic damage was reported for different parts of the central nervous system (6,11), but also for different layers of the cerebral cortex with the cells of the third layer being the most sensitive (3). However, also various cells of a layer and even various neurons of one type can be differently vulnerable by ischemia (11). This different susceptibility of individual cells to impaired blood flow became obvious in our study, but its cause remained unclear partly because the cells recorded from could not be identified. Some cells may tolerate a more severe flow reduction due to their location close to a supplying blood vessel where oxygen tension and substrate consumption is relatively high in spite of low regional perfusion. In other neurons, the differences in ischemic tolerance may be due to specific cell properties, especially their size: it was indicated that cells recorded as action potentials of small amplitudes had a lower functional threshold. It may be argued that small extracellular spikes are caused by small electric fields originating from small cells. Therefore, it might be indicated in our results that larger cells, i.e. pyramidal cells of the fifth and third layer, are more sensitive to ischemic damage than small neurons of the other layers. However, the function of a neuronal network is impaired when the most sensitive part is damaged. The ischemic functional and morphological threshold of this most sensitive neuron is therefore critical for reversible or irreversible damage of neuronal function.

References

1. Aukland K, Bower BF, Berliner RW (1964) Measurement of local blood flow with hydrogen gas. Circ Res 14:164-187

2. Branston NM, Symon L, Crockard HA, Pasztor E (1964) Relationship between the cortical evoked potential and local cortical blood flow following acute middle cerebral artery occlusion in the baboon. Exp Neurol 45:195-208

3. Brierley JB (1976) Cerebral hypoxia. In: Blackwood W, Corsellis JAN. Eds. Greenfield's Neuropathology. Edward Arnold, London, pp 43-85

4. Heiss W-D, Hayakawa T, Waltz AG (1976) Cortical neuronal function during ischemia. Arch Neurol 33:813-820

5. Heiss W-D, Rosner G (1982) Duration versus severity of ischemia as critical factors of cortical cell damage. 13th Princeton Conference Cerebrovascular Diseases, in press

6. Heymans C, Boukaert JJ (1935) Sur la vie et la reanimation des centres nerveux. Compt rend Soc de Biol 119:324

7. Jones TH, Morawetz RB, Crowell RM, Marcoux FW, Fitzgibbon SJ, DeGirolami U, Ojemann RG (1981) Thresholds of focal cerebral ischemia in awake monkeys. J Neurosurg 54:773-782

8.   Little JR (1977) Implanted device for middle cerebral artery occlusion in conscious cats. Stroke 8:258-260

9.   Morawetz RB, DeGirolami U, Ojemann RG, Marcoux FW, Crowell RM (1978) Cerebral blood flow determined by hydrogen clearance during middle cerebral artery occlusion in unanesthetized monkeys. Stroke 9:143-149

10. Rappelsberger P, Heiss W-D, Volmer R, Turnheim M (1977) A technique for recording local blood flow and neuronal activity with a single microelectrode. Pflügers Arch 369:183-186

11. Scholz W (1953) Selective neuronal necrosis and its topistic patterns in hypoxemia and oligemia. J Neuropath Exp Neurol 12:249-261

12. Sharbrough FW, Messick JM, Sundt TM jr (1973) Correlation of continuous electro-encephalograms with cerebral blood flow measurements during carotid end-arterectomy. Stroke 4:674-683

13. Spielmeyer W (1922) Histopathologie des Nervensystems. Springer, Berlin

14. Traupe H, Heiss W-D, Umbach C (1981) Microflow and cortical neuronal function during temporary ischemia. J Cereb Blood Flow Metabol 1, Suppl 1:190-191

15. Trojaborg W, Boysen G (1973) Relation between EEG, regional cerebral blood flow and internal carotid artery pressure during carotide endarterectomy. Electroenc Clin Neurophysiol 34:61-69

16. Umbach D, Heiss W-D, Traupe H (1981) Effect of graded ischemia on functional coupling and components of somatosensory evoked potentials. J Cereb Blood Flow Metab 1, Suppl 1:198-199

Membrane Stabilization and Protection of the Ischemic Brain

J. Astrup

Department of Neurosurgery, Rigshospitalet, Blegdamsvey 9, 2100 Copenhagen, Denmark

If cerebral circulation is arrested, the oxidative metabolism and hence almost all ATP production will cease. The cells rapidly become depleted of ATP, and the energy requiring processes, mainly active ion transport, are arrested. The cells leak K^+ and take up Na^+, Cl^-, Ca^{2+} and water, and the membranes depolarize. This is the general scheme leading to the terminal stage of ischemic membrane failure. It is obeyed by all tissues subjected to acute ischemia, but the remarkable feature of the brain is the rapidity with which this occurs. Due to extreme leakiness of the cellular compartment in the brain, the state of terminal ischemic membrane failure is reached in a matter of minutes, while it may take hours or even days for the passive ion leak fluxed to come to a stop in other organs with tight membranes such a muscle, peripheral nerve, or erythrocytes.

The high rate of Na^+ and K^+ leak fluxes in the brain when subjected to ischemia suggests profound membrane leakiness and a high rate of leak fluxes of these ions even in normal nonischemic brain. These cation leak fluxes must of course be continuously counteracted by ATP-consuming Na^+-K^+ transport. This suggests that "membrane stabilization" which in the present context is defined as a block of the Na^+-K^+ leak fluxes, has two effects: one is to slow the cellular leak of K^+ and Na^+ in the ischemic brain, and the other is to reduce ATP consumption and hence oxygen and glucose uptake in normal nonischemic brain. Lidocaine blocks the Na^+ channels in peripheral nerves. As will be summarized in this presentation, we have found that this drug has a similar action in the brain. Lidocaine in high doses effectively reduces the cellular leak of K^+ in the ischemic brain, and in the nonischemic brain the drug reduces oxygen and glucose consumption. Hypothermia has a similar effect and it appeared to be additive to that of lidocaine. The data leading to these conclusions have been fully accounted for elsewhere (2,4), and will briefly be reviewed here, with specific emphasis on prospects of protection of the ischemic brain.

Methods and experimental design

Dogs were used as experimental animals. In one series, the extracellular K^+ concentration in the brain cortex was monitored by a surface potassium electrode (4) and the effect of thiopental 40 mg/kg and of lidocaine 160 mg/kg on the rate of rise in the surface K^+ concentration following circulatory arrest was studied at brain temperatures of 37, 28 and 18°C. In a second series, cerebral blood flow and oxygen and glucose uptake ($CMRO_2$ and CMRgl) were measured by the sagittal sinus outflow method of Rapela et al. (13) and Michenfelder et al. (10). A stepwise experimental procedure followed. First, the synaptic transmission and associated metabolism was blocked by barbiturate, and second, the assumed high rate of leak fluxes remaining in the functional inactivated brain with flat EEG was blocked by lidocaine. Studies were carried out at brain temperatures of 37, 28 and 18°C.

In brief, methods included cardiopulmonary bypass circulation allowing deep hypothermia and well defined circulatory arrest periods and recirculation. Systemic BP was maintained above 50 torr by adjusting the pump speed. Gas flow through the oxygenator was kept constant during cooling. This allowed arterial pCO_2 to fall with the diminished CO_2 production. Experiments ran to normoglycemia. Surgical anesthesia was maintained during by-pass circulation by halothane 1% and nitrous oxide 25% in atmospheric air.

Results

Extracellular potassium concentration $(K^+)_s$

The increase in the brain surface K^+ concentration $(K^+)_s$, indicating the cellular leak of K^+ during circulatory arrest, was unaffected by thiopental 40 mg/kg, but was slowed by lidocaine 160 mg/kg and hypothermia 28 and 18°C. An additive effect of lidocaine and hypothermia was observed. Figure 1 includes the potassium electrode traces from each single arrest episode in all the experiments, the control condition being halothane-nitrous oxide anesthesia. The rise in $(K^+)_s$ to e.g. 20 mM was reduced from 4 - 5 min at 37°C and no drugs to about 30 min at 18°C in combination with lidocaine.

Cerebral blood flow and metabolims

The influence of brain temperature on cerebral metabolism is well described and shall not be further outlined here. Like Michenfelder and Theye (11), we found an Arrhenius type of interrelation having a Q_{10} of 2.47.

More interest is accorded to the effects of pentobarbital and lidocaine. When studied separately, both drugs caused flat EEG in the dose given, and this effect was accompanied by profound metabolic depression. In percent of control, this depression was equally pronounced at 37, 28 and 18°C. When studied in combination, it was found that lidocaine, when given after pentobarbital (flat EEG), caused additional metabolic depression of about 15 - 20%, again equally pronounced at the three temperature levels. On the contrary, when pentobarbital was given after lidocaine (flat EEG), no additional metabolic depression was observed. The data is summarized in the table.

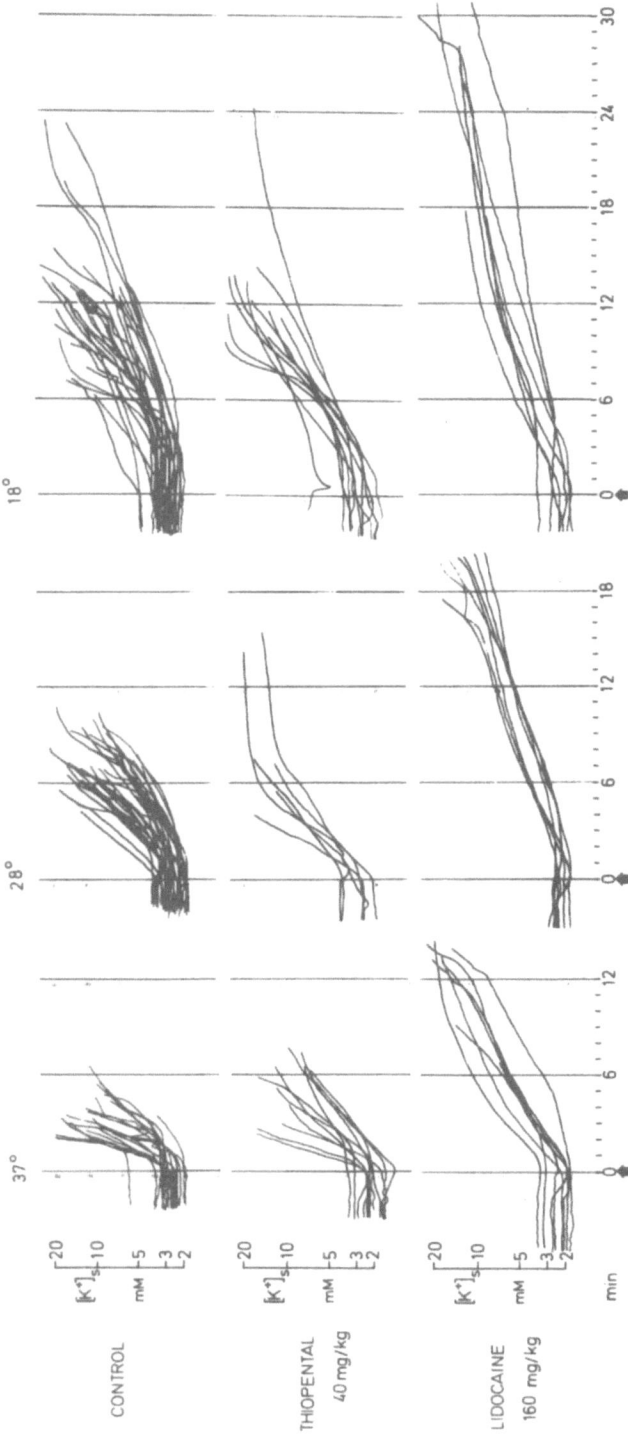

Fig. 1   Potassium efflux curves grouped according to brain temperature and drug. The length of the curves indicate circulatory arrest duration, since the recirculation phase is left out

From Astrup et al. Anesthesiology 55: 255–262, 1981

34

Table 1   Effect of pentobarbital and lidocaine on cerebral oxygen and glucose consumption. The separate effects of the drugs are shown in section 1, and the combined effects in section 2 and section 3. Values are means $\pm$ SEM

| | Section 1 | | | | |
| | CMR_{O_2} percent | | | CMR_{gluc} percent | |
	37°C	28°C	18°C	37°C	28°C
Halothane 1-1.5 percent control	100	100	100	100	100
Pentobarbital (40 mg/kg)					
Mean	69.4	69.6	74.7	74.2	87.2
SEM	5.5	-	-	6.1	-
n	6	3	3	6	3
Lidocaine (160 mg/kg)					
Mean	65.1	71.0	58.0	61.2	73.2
SEM	4.7	-	-	5.1	-
n	6	4	2	6	2
	Section 2				
Pentobarbital (40 mg/kg)	100	100	100	100	100
Lidocaine (160 mg/kg)					
Mean	84.8	84.7	87.7	81.0	72.0
SEM	3.4	-	-	5.3	-
n	6	3	3	6	3
	Section 3				
Lidocaine (160 mg/kg)	100	100	100	100	100
Pentobarbital (40 mg/kg)					
Mean	95.8	104.5	105	101.3	115
SEM	-	-	-	-	-
n	3	2	1	3	2

From: Astrup et al. (1981) Anesthesiology 55:263-268

Discussion

We interpret the results outlined above in the following way. Pentobarbital (or thiopental) specifically blocks synaptic transmission (flat EEG) and inhibits connected energy requiring cell functions, mainly Na^+-K^+ transport at the synaptic sites, but has no detectable effect on membrane permeability for these ions and hence no effect on the Na^+-K^+ leak fluxes or connected ion transport and energy consumption (no slowing of the ischemic K^+ efflux, no additional metabolic depression after lidocaine). Lidocaine has a dual action. The drug blocks synaptic transmission and connected metabolism (a "barbiturate-like" effect), and in addition to this the drug causes a specific block of the Na^+ channels in the membranes (slowing of the ischemic K^+ efflux, additional metabolic depression after pentobarbital and flat EEG). This is the so-called additional "membrane stabilizing" effect of lidocaine. It should be noted here that lidocaine was given in a dose sufficiently high to avoid the seizure-eliciting effect of intermediate doses (14).

Hypothermia causes progressive flattening of the EEG (a "barbiturate-like" effect), and slowing of the ischemic K^+ efflux (a "membrane stabilizing" effect), and these and probably others slowing effects are associated with profound metabolic depression.

These results in combination with additional studies using ouabain as experimental tool to inhibit Na^+-K^+ transport have led to the conclusion that total energy consumption in the brain falls into three categories (5). One is connected to intensity of synaptic transmission and varies with brain work, one is connected to the Na^+-K^+ leak fluxes and its counteracting

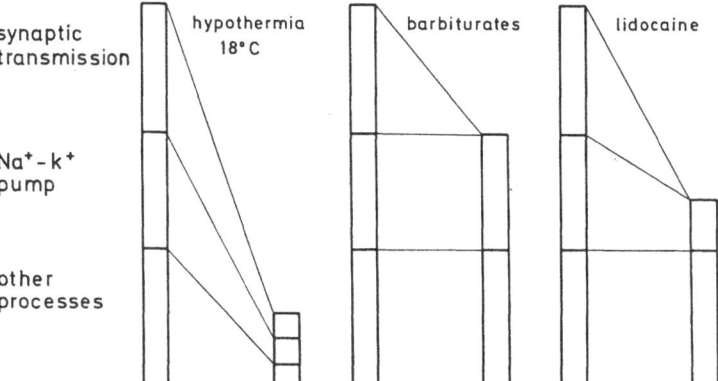

Fig. 2   Schematic representation of cerebral function and associated metabolism. Inhibition by hypothermia, barbiturates, and lidocaine. See explanation in text

Na^+-K^+ transport, and the last part is associated other as yet incompletely defined processes such as Ca^{2+} transport, transmitter and protein turnover, axoplasmic transport etc. This scheme of cerebral energy consumption is outlined in Fig. 2. It further indicates the proposed effects of barbiturate, lidocaine and hypothermia as metabolic depressants.

Possible implications for the problem of protection of the brain during circulatory arrest

There is evidence that the irreversible structural ischemic damage, irrespective of its much debated underlying mechanism, are "uncovered" or "triggered" by the appearance of the ischemic state of membrane failure, i.e. ATP depletion, arrested active transport, net leak fluxes of Na^+, K^+ and Ca^{2+} and membrane depolarization. The evidence for this mainly derives from studies of middle cerebral artery occlusion and focal incomplete ischemia. In areas of critical low collateral blood flow there seems to be a correlation between the appearance of ischemic membrane failure (ATP depletion, leak of cellular K^+ etc. leading to membrane depolarization) and the development of irreversible damage of tissue structure and function (3,6,9,12,16,17).

Accordingly, one mode of protection is to prevent or at least to delay the ischemic membrane failure as much as possible as indicated i.e. by slowing of the ischemic K efflux. In case of focal ischemia, the ischemic K^+ efflux can be prevented, at least in the acute state, by supporting local blood flow above the critical low threshold below which the active Na^+-K^+ transport fails. In case of complete global ischemia as circulatory arrest, the ischemic K^+ efflux cannot be prevented, only delayed. Hypothermia as well as the state of immaturity of the brain represent two conditions of slowed ischemic K^+ efflux, and both these conditions are associated with marked clinical protection against ischemia/hypoxia. This had led to reconsideration of the hypothesis first suggested by Bures and Buresova (7), stating that one mode of protection of the brain during circulatory arrest is to delay the ischemic membrane failure. Accordingly, it has been suggested that lidocaine (which causes a delay of the ischemic membrane failure) may induce a similar delay in the development of ischemic infarction (1,2,4,5). It is of interest that the effect of lidocaine seems to be additive to the effect of hypothermia. This suggests the possibility of additive protection by this combination. Such a combined effect may prove useful in selected clinical cases, e.g. within the fields of cardiac surgery and neurosurgery, occasionally requiring prolonged circulatory arrest under deep hypothermia protection for the surgical procedure (8,15).

References

1. Astrup J (1982) Energy-requiring cell functions in the ischemic brain. J Neurosurg 56:482-497

2. Astrup J, Møller Sørensen P, Rahbek Sørensen H (1981) Inhibition of cerebral oxygen and glucose consumption in the dog by hypothermia, pentobarbital, and lidocaine. Anesthesiology 55:263-268

3. Astrup J, Siesjö BK, Symon L (1981) Thresholds in cerebral ischemia -the ischemic penumbra (editorial). Stroke 8:51-57

4. Astrup J, Skovsted P, Gjerris F, Rahbek Sørensen H (1981) Increase in extracellular potassium in the brain during circulatory arrest: Effects of hypothermia, lidocaine, and thiopental. Anesthesiology 55:256-262

5. Astrup J, Sørensen PM, Rahbek Sørensen H (1981) Oxygen and glucose consumption related to Na-K transport in canine brain. Stroke 12:726-730

6. Astrup J, Symon L, Branston NM, Lassen NA (1977) Cortical evoked potential and extracellular K^+ and H^+ at critical levels of brain ischemia. Stroke 8:51-57

7. Bures J, Buresova O (1975) Die anoxische Terminaldepolarisation als Indikator der Vulnerabilität der Großhirnrinde bei Anoxie und Ischämie. Pflügers Arch 264:325-334

8. Dillard DH, Mohri H, Merendino KA (1971) Correction of heart disease in infancy utilizing deep hypothermia and total circulatory arrest. J Thorac Cardiovasc Surg 61:64-69

9. Jones TH, Morawetz RB, Crowell RM (1981) Thresholds of focal cerebral ischemia in awake monkeys. J Neurosurg 54:773-782

10. Michenfelder JD, Messick JM, Theye RA (1968) Simultaneous cerebral blood flow measured by direct and indirect methods. J Surg Res 8:475-481

11. Michenfelder JD, Theye RA (1968) Hypothermia: Effect on canine brain and whole-body metabolism. Anesthesiology 29:1107-1112

12. Morawetz RB, De Girolami U, Ojemann RG (1978) Cerebral blood flow determined by hydrogen clearance during middle cerebral artery occlusion in unanesthetized monkeys. Stroke 9:143-149

13. Rapela CE, Green HD, Denison AB (1967) Baroreceptor reflexes and autoregulation of cerebral blood flow in the dog. Circ Res 21:559-568

14. Sakabe T, Maekawa T (1974) The effects of lidocaine on canine cerebral metabolism and circulation related to the electroencephalogram. Anesthesiology 40:433-441

15. Silverberg GD, Reitz BA, Ream AK (1981) Hypothermia and cardiac arrest in the treatment of giant aneurysms of the cerebral circulation and hemangioblastoma of the medulla. J Neurosurg 55:337-346

16. Symon L, Crockard HA, Dorsch NWC (1975) Local cerebral blood flow and vascular reactivity in chronic stable stroke in baboons. Stroke 6:482-492

17. Tamura A, Asano T, Sano K (1980) Correlation between CBF and histological changes following temporal middle cerebral artery occlusion. Stroke 11:487-493

Recovery From Disturbed Cerebral Ion Homeostasis Following Severe Incomplete Ischemia and Modification by the Metabolic Depressant Drug Etomidate

D. Heuser and H. Guggenberger

Abteilung für Anästhesiologie, Universität Tübingen, Calwerstrasse 7, 7400 Tübingen, FRG

Therapeutic management with metabolic depressant drugs (e.g. barbiturates, althesin, etomidate) in patients suffering from major disturbances in cerebral blood flow (CBF) and metabolism has become increasingly accepted. Results from animal studies with different approaches to cerebral ischemic lesion (11,23,24) as well as from patients with severe head injuries (9,27) have indicated that pharmacologically induced "low flow and low metabolism" do benefit to the brain by preventing a mismatch between oxygen demand and delivery under critical flow conditions with subsequent lactacidosis and membrane failure; such conditions are regarded as the most important precursors of irreversible cell damage (2,21). Despite numerous promising results under regional and incomplete global cerebral ischemia (1,13), it must be admitted that the majority of the results subsequent to global interruption of cerebral circulation are unfavorable. Consequently, metabolic depression may only represent one part of the ideal therapeutic regimen and may be ineffective in those cases in which functional activity is already abolished (10). It has been suggested by several authors that the beneficial effects of these drugs under the above mentioned conditions are not only due to depression of oxidative metabolism, but also to other peculiar effects which are characteristic of each drug. This hypothesis has been supported by the work of Carlsson and Rehncrona (3) who demonstrated that chlormethiazole does not exert protective effects on cerebral postischemic recovery, but reduces CBF and metabolism to the same degree as barbiturates. In the case of barbiturates, the specific effects could be attenuation of free fatty acid liberation (15), free radical scavenging (15), amelioration of low CBF in poorly perfused areas via an inverse steal phenomenon (4), improvement of microcirculation by decreasing intracranial pressure (20) and prevention of cerebral edema (22).

Since ionic gradients are important preconditions for physiological function of all excitable tissues, one could imagine that ions are also basically involved in the functional recovery process of central nervous structures after ischemic injury. Aim of the present animal study was to analyze whether the recovery of disturbed extracellular ion homeostasis following severe incomplete cerebral ischemia could be ameliorated by metabolic depressant drugs, thus offering one explanation of better outcome of patients treated with such drugs under similar clinical conditions.

Methods and experimental procedures

Experiments were performed with 21 cats (2.1 - 3.6 kg body weight) subjected to general anesthesia with halothane (0.5 - 1.5 Vol%) under relaxation (pancuronium bromide 0.2 mg/kg) and mechanical ventilation. After establishing two arterial and venous lines via the femoral vessels for continuous monitoring of mean arterial blood pressure (MABP),   blood sampling and administration of maintenance fluids (3 ml/kg/hour), the animal was brought into the sphinx position, the head fixed in a stereotaxic device. Craniotomy was then performed on both sides in the fronto-parietal region and, after careful removal of dura mater, an area of cerebral cortex (1 x 1 cm) was exposed. Under stereomicroscopic observation, pH and K^+ microelectrodes were implanted 200 - 300 μm into the cortical tissue together with a reference electrode which consisted of a chlorinated silver wire in a micropipette filled with physiological saline in 1 % agar. A second reference of the same composition but with larger dimension was in contact with the animal's blood and permitted continuous registration of shifts in dc-potential. The ion-selective devices were constructed using well-established techniques (6) and calibrated in appropriate buffer solutions as described earlier (12). Local cerebral blood flow was measured using the hydrogen inhalation clearance method (25) and platinum wires of 30 - 60 μm diameter for implantation. On both hemispheres, EEG was registered with two electrodes alligned in the fronto-occipital direction at a distance of 3 - 4 mm from the ion-selective devices. As indicated by the numbers in the experimental protocol (Fig. 1), blood samples were taken at certain intervals

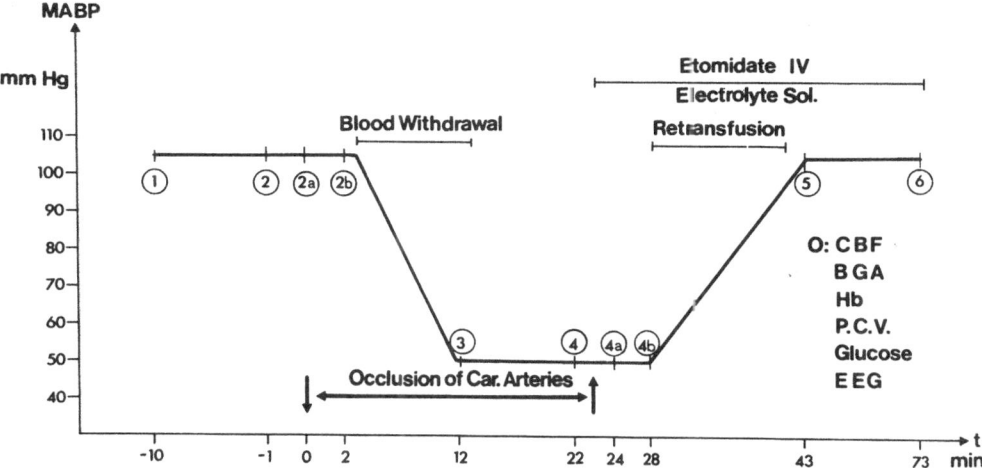

Fig. 1   Experimental procedure for induction of incomplete cerebral ischemia in 21 cats by bilateral carotid artery occlusion and blood withdrawal. The numbers in the open circles indicate analysis of cerebral blood flow (CBF), blood gases (BGA), hemoglobin (Hb), packed cell volume (PCV), glucose, electrical activity EEG

to verify changes in acid base balance, hemoglobin, glucose and hematocrit. Additional measurements of rectal temperature as well as of endtidal CO_2 helped to establish physiological control conditions. Critical levels of cerebral blood flow were established using the combination of carotid clamping and blood withdrawal. Throughout incomplete ischemia MABP was maintained at 50 mmHg by withdrawal or reinfusion of the animal's blood. Afterwards, clamping was released and blood was reinfused in order to reestablish preischemic MABP conditions. In 11 animals, etomidate (5 mg/kg i.v. initially, followed by 4 mg/kg/h) was administered concomitantly to reperfusion, whereas ten control animals were merely treated with physiological saline. Measurements of the individual parameters were continued to the point of normalization of ionic gradients up to at least 30 minutes after reperfusion. Statistical evaluation of the results was performed using the Wilcoxon-Pratt test.

Results

As indicated in Table 1, there were no statistically significant differences in essential physiological parameters of the two groups at the beginning of the experiments and during incomplete cerebral ischemia. As already known (7), severe incomplete ischemia results in distinct extracellular acidosis, which is more pronounced than during complete ischemia and does not seem to reach a plateau because of maintained anaerobic glycolysis during residual cerebral glucose availability. If CBF decreases below 10 - 15 ml/100 g/min, extracellular K^+ increases, indicating membrane failure and subsequent structural damage if this state persists. During the reperfusion period, extracellular ion activities normalized significantly more quickly in the control group in comparison to the etomidate-treated animals (Table 2). In both groups, the immediate recirculation phase was characterized by transient enhancement of tissue acidosis before pH_e began to normalize. However, this further decrease in pH_e was significant only in the etomidate-treated group.

Extracellular potassium increased in cases of flow reduction below 10 - 15 ml/100 g/min. During the initial period of recirculation, extracellular K^+ seemed to normalize more quickly in the etomidate group, however, the second phase, culminating in normal K^+_e, was significantly prolonged (Table 1) in comparison to the control group. Local CBF was significantly higher in the control group during the postischemic reperfusion period (105 ml/100 g/min VS 68.5 ml/100 g/min), and increased in both groups in comparison to initial values. The maximum of postischemic hyperperfusion was observed significantly later in the etomidate-treated group. Changes in electrical activity were similar; recovery was complete in five (out of ten) animals of the control group, whereas six (out of 11) cats in the etomidate group revealed reappearance or maintenance of electrical function. As expected, spontaneous EEG did not appear until extracellular K^+ was within the physiological range, but was not present in all cases of reattained physiological ion homeostasis.

Table 1 Control and extreme values attained prior to and
        at the end of severe incomplete cerebral ische-
        mia. Mean arterial blood pressure (MABP) and ex-
        tracellular pH (pH_e) and potassium (K_e^+) were
        measured

	Etomidate Group		Control Group	
	before inc.	end of ischemia	before inc.	end of ischemia
MABP (mmHg)	118	50	103	46
pH_e	7.301	5.815	7.255	6.282
K_e^+ (mM)	4.9	51.8	4.85	49.8

Table 2 Time to normalization of cerebral ion homeostasis
        in cats after 15 min severe incomplete cerebral
        ischemia. Reattainment of ionic equilibrium is
        significantly (*) delayed in the etomidate-trea-
        ted group

	Etomidate Group	Control Group
pH_e normalization (min)	131.5	60.5
K_e^+ normalization (min)	53	31

Discussion and conclusions

The main result of the present study is that normalization of disturbed ionic homeostasis in the brain subsequent to severe incomplete ischemia is delayed by postischemic treatment with high doses of etomidate. In the case of pH_e, this has already been suggested by Nemoto and coworkers (14) in complete interruption of cerebral circulation, but could not be verified statistically. The question remains, however, as to whether this attenuation could be of any beneficial effect to the outcome of patients with similar clinical conditions. It would seem reasonable to assume that the most rapid normalization of disturbed ion homeostasis is desirable since this is an important precondition for normal electrical function of central nervous structures. Furthermore, the present results clearly indicate that this recovery is due to adequate postischemic perfusion, since mean flow values in the control group were nearly twice as high (105 ml/100 g/min VS 68.5 ml/100 g/min) as in the etomidate-treated group. Nevertheless, one should consider that postischemic hyperperfusion is accompanied by cytotoxic edema formation (28) which can (e.g. by raising intracranial pressure) be followed by a secondary ischemic event, which, in turn, may easily outweigh the initial quick normalization of ion homeostasis. This potential source of interference could not be measured using the present open skull preparation, however. In light of the additional, specific, beneficial effects which have been reported by other groups and have been mentioned above, it would seem worthwhile to use metabolic depressant drugs in the anesthetic management of severe incomplete cerebral ischemia. Sufficiently moderate dosages are required, however, in order to provide adequate perfusion pressure without cardiovascular side effects and on the other hand to inhibit the local steal phenomena which are nearly obligatory under conditions of pronounced tissue hyperperfusion. It is not yet clear whether etomidate also possesses those above-mentioned specific properties which could ameliorate the sequelae of ischemic encephalopathy. However, recent experimental data (26) support the efficacy of the drug.

The enhancement of extracellular acidosis directly after release of clamping and the beginning of reperfusion, which was significantly more pronounced in the etomidate group, could be due to increased lactate production by activated glycolysis in the early period of flow improvement (18). Since the clearance of lactate from the brain is also attenuated (19), such further decrease in pH_e can be expected. Inhibition of oxidative metabolism by etomidate under these conditions retards oxidation of lactate (17) and may thus explain the present significant differences in pH_e in the early post-ischemic phase. The more rapid restoration of K^+_e in the immediate phase of recirculation in the etomidate group (in contrast to the delay in attaining the initial value), however, is not easy to explain. In the case of metabolic inhibition, one could discuss whether the ATP generated by recirculation primarily used to reestablish disturbed structures and membrane gradients, thus making "better use" of ATP which otherwise would have required for other purposes. Such a notion presumes a certain compartmentalization of energy-rich compounds. It is quite clear that the present study can only make a limited contribution to the whole problem of cerebral protection, especially since consideration has not been given to the problems of the late postischemic period, when cerebral hypoperfusion is confronted with increased metabolism due to functional recovery. Under such conditions, flow cannot be enhanced by increasing perfusion pressure since autoregulation is working again, but in the absence of CO_2 reactivity (16). The resulting mismatch between oxygen demand and supply would certainly again result in severe disturbances of cerebral ion homeostasis which could be prevented or at least attenuated by treatment with metabolic depressant drugs, as indicated by experiments currently underway in our laboratory. Whether other therapeutic approaches (e.g. calcium antagonists, antianoxic drugs etc.) may be more promising is not yet clear; their possible benefits have been discussed in a recent review (8).

Acknowledgement

The study was supported by a grant from the Deutsche Forschungsgemeinschaft (He 1101/3-3).

References

1. Astrup J (1980) Barbiturate protection in focal cerebral ischemia. Scand J Clin Lab Invest 40:201-203

2. Branston NM, Strong AJ, Symon L (1977) Extracellular potassium activity, evoked potential and tissue blood flow. J Neurol Sci 32:305-321

3. Carlsson C, Rehncrona S (1979) Influence of chlormethiazole on cerebral blood flow and oxygen consumption in the rat, and its effect on the recovery of cortical energy metabolism after pronounced incomplete ischemia. Acta Anaesthesiol Scand 23:259-266

4. Feustel PJ, Ingvar MC, Severinghaus JW (1981) Cerebral oxygen availability and blood flow during middle cerebral artery occlusion: Effects of pentobarbital. Stroke 12:858-863

5. Flamm ES, Demopoulos HB, Seligman ML, Poser RG, Ransohoff J (1978) Free radicals in cerebral ischemia. Stroke 9:445-447

6. Heuser D (1981) Local ionic control of cerebral microvessels. In: Zeuthen T. Ed. The application of ion-selective microelectrodes. Elsevier/North Holland, pp 85-105

7. Heuser D, Morris PJ, McDowall DG (1981) Ionic changes in the brain with ischemia. In: Zindler M, Rügheimer E. Eds. Anaesthesiology; Proceed 7th World Congr Anaesthesiologists, Hamburg 1980. Excerpta Medica, Amsterdam Oxford Princeton, pp 821-826

8. Hossmann KA (1982) Treatment of experimental cerebral ischemia. J Cereb Blood Flow Metabol 2:275-297

9. Marshall LF, Smith RW, Shapiro HM (1979) The outcome with aggressive treatment in severe head injury. J Neurosurg 50:26-30

10. Michenfelder JD (1974) The interdependency of cerebral functional and metabolic effects following massive doses of thiopental in the dog. Anesthesiology 41:231-236

11. Michenfelder JD, Milde JH (1975) Influence of anesthetics on metabolic, functional and pathological responses to regional cerebral ischemia. Stroke 6:405-410

12. Morris PJ, Heuser D, McDowall DG, Hashiba M (1982) Extracellular pH and potassium activity in the cerebral cortex during hypotension induced with trimetaphan or sodium nitroprusside. Anesthesiology, in press

13. Moseley JI, Laurent JP, Molinari GF (1975) Barbiturate attenuation of the clinical course and pathologic lesions in a primate stroke model. Neurology 25:870-874

14. Nemoto EM, Frinak S (1981) Brain tissue pH after global brain ischemia and barbiturate loading in rats. Stroke 12:77-82

15. Nemoto EM, Shiu G, Alexander H (1980) Brain free fatty acids (FFA) during decapitation ischemia in awake and pentobarbital anesthetized rats. Fed Proc 39:407

16. Nemoto EM, Snyder JV, Carroll RG, Morita H (1975) Global ischemia in dogs: Cerebrovascular CO_2 reactivity and autoregulation. Stroke 6:425-431

17. Nordström CH, Rehncrona S (1977) Postischemic cerebral blood flow and oxygen utilization in rats anesthetized with nitrous oxide or phenobarbital. Acta Physiol Scand 101:230-240

18. Nordström CH, Rehncrona S, Siesjö BK (1978) Restitution of cerebral energy state, as well as of glycolytic metabolites, citric acid cycle intermediates and associated amino acids after 30 minutes of complete ischemia in rats anesthetized with nitrous oxide or phenobarbital. J Neurochem 30:479-486

19. Oldendorf W, Braun L, Cornford E (1979) pH dependence of blood-brain barrier permeability to lactate and nicotine. Stroke 10:577-581

20. Shapiro HM, Wyte StR, Loeser J (1974) Barbiturate-augmented hypothermia for reduction of persistent intracranial hypertension. J Neurosurg 40:90-100

21. Siesjö BK (1981) Cell damage in the brain: A speculative synthesis. J Cereb Blood Flow Metabol 1:155-185

22   Simeone FA, Frazer G, Lawner P (1979) Ischemic brain edema: Comparative effects of barbiturates and hypothermia. Stroke 10:8-12

23. Smith AL, Hoff JT, Nielsen SL, Larson CP (1974) Barbiturate protection in acute focal cerebral ischemia. Stroke 5:1-7

24. Steen PA, Michenfelder JD (1978) Cerebral protection with barbiturates: Relation to anesthetic effect. Stroke 9:140-142

25. Symon L, Pasztor E, Branston NM (1974) The distribution and density of reduced cerebral blood flow following acute MCA occlusion: An experimental study by the technique of hydrogen clearance in baboons. Stroke 5:355-364

26. Wauquier A (1982) Brain protective properties of etomidate and flunarizine. J Cereb Blood Flow Metabol 2, Suppl. 1:S53-56

27. Wiedemann K, Hamer J, Weinhardt F, Just OH (1980) Barbituratinfusion bei schwerem Schädelhirntrauma. Anaesth Intensivther Notfallmed 15:303-314

28. Zimmermann V, Hossmann V, Hossmann KA (1975) Intracranial pressure after prolonged cerebral ischemia. In: Lundberg N, Ponten U, Brock M. Eds. Intracranial Pressure II. Springer, Berlin, pp 177-182

Comparative Evaluation of Barbiturate and CA^{++} Antagonist Attenuation of Brain Free Fatty Acid Liberation During Global Brain Ischemia

G. K. Shiu, E. M. Nemoto, J. P. Nemmer, and P. M. Winter

The Anesthesia and CCM Research Laboratories, Department of Anesthesiology and Critical Care Medicine, University of Pittsburgh, School of Medicine, 1081 Scaife Hall, Pittsburgh, PA 15261, USA

Introduction

Whole brain fatty acids (FFA) increase rapidly and continue to rise even after prolonged durations of complete global ischemia with a time course and magnitude unlike that of any other cerebral metabolite known thus far (4,5). Therefore, we hypothesized that brain FFA accumulation may reflect the evolution of ischemic brain injury at least during complete global ischemia. If so, drugs effective in improving the tolerance of the brain to ischemic anoxic insults such as the barbiturates (1,2) should attenuate the liberation of FFAs. Accordingly, pentobarbital anesthesia signficantly attenuated whole brain FFA liberation with more pronounced effects as the duration of ischemia was prolonged (4,5). This finding provided the first evidence of a metabolic basis for barbiturate-improved tolerance of the brain to complete global ischemia.

If the attentuation of FFA liberation by a drug indicates its efficacy in attentuating ischemic brain damage, it may be used as a means of evaluating the potential efficacy of various drugs. Thus, we did a comparative evaluation of several drugs of varying reported efficacy in ameliorating ischemic brain injury after focal and global ischemia (3). We found that in general, those drugs most widely documented in their ability to reduce the severity of ischemic brain damage attenuated whole brain FFA most dramatically and especially on arachidonic acid liberation. The drugs we studied fell into the following three groups in order of decreasing effectiveness in attenuating FFA accumulation after 10 min decapitation global ischemia in rats (3): 1) Phenytoin, thiopental, pentobarbital, innovar, 2) R41-468, Y-9178, etomidate, and 3) lofentanil, halothane and ketamine. The relative effectiveness of the various drugs were most demonstrable in terms of arachidonic acid.

Aside from the question of which drugs are most effective in ameliorating ischemic anoxic insults, for the clinical application of the drugs, it is very important to know the minimum dose needed to produce maximum therapeutic effects especially for those drugs that may be potent respiratory and cardiovascular depressants. Pursuing the same line of reasoning on FFA accumulation during global ischemia, the degree of attentuation by a given drug at various doses may provide some insight into the optimum dose for therapeutic effects.

We therefore studied the relationships in FFA attenuation for effective drugs such as phenytoin, thiopental and pentobarbital and compared them with an ineffective drug, ketamine (6). The data showed that maximal attenuation of FFA accumulation after 10 min decapitation ischemia in rats occurred at pentobarbital and thiopental doses equal to or less than that required for surgical anesthesia. Maximal attentuation by phenytoin occurred at the anticonvulsant dose. Ketamine, a reportedly ineffective drug at least relative to pentobarbital in focal ischemia, reduced total FFA at a dose of 150 mg/kg but more importantly, was totally ineffective in attenuating arachidonic acid liberation. These results have several important clinical and mechanistic implications. Clinically, they suggest that for maximal therapeutic effects, anesthetic or subanesthetic doses of the barbiturates are adequate which is extremely important since these drugs are potent cardiovascular and respiratory depressants. Mechanistically, the data indicate that depression of cerebral metabolism is not essential for maximum therapeutic effects and is probably not the primary mechanism of action since phenytoin is equally if not more effective than the barbiturates, but depresses cerebral metabolism minimally. Thus, the unproven premise that for maximal therapy, the EEG should be suppressed to the isoelectric state may not only be untrue, but also dangerous.

Before progressing further on this theme, it is important to emphasize the limitations of the model we are using to evaluate the efficacy of pharmacotherapy in cerebral ischemic anoxic insults. First, we can only administer the drugs preinsult which is not the clinical circumstance. However, the fundamental mechanisms, we believe, in cell injury at least at the cellular level following ischemic anoxia are similar whether it occurs during or after the insult. The factors initiating the insult may differ after recirculation following cardiac arrest from that occurring during the arrest but the end result, namely, energy depletion of the cell and inability to maintain intracellular ionic homeostasis is probably the same.

Second, only those therapies that act directly on the metabolic processes occurring during or after ischemic insult can be evaluated. For example, those therapies that may improve neurologic recovery by improving perfusion after resuscitation such as hemodilution and osmotherapy would not be effective in our model. Furthermore, those drugs that may improve circulation postinsult by altering vasomotor tone and therefore perfusion such as the calcium antagonists, would not necessarily be revealed as effective agents in our model unless they also have an effect on cerebral metabolism independent of their cerebrovascular effects.

These limitations in the model do not, however, preclude any statements on the possible beneficial effects of a drug when administered postinsult on the basis of their effects preinsult. However, the Ca^{++} antagonists may effectively inhibit FFA liberation during ischemia and especially arachidonic acid which as a precursor to prostaglandins, may have detrimental effects on cerebral circulation and metabolism postischemia. Thus, the attenuation of arachidonic acid liberation during ischemia is of consequence after recirculation.

Recently, there has been a great deal of interest in the applicability of the calcium antagonists in cerebral resuscitation. The possible mechanisms whereby Ca^{++} antagonist may be of benefit in attenuating postischemic encephalopathy is as numerous as the roles of calcium in physiological and biochemical processes ranging from muscle contraction to synaptic transmission and neurotransmitter release. A well-recognized role of Ca^{++} is its activation of phospholipases. Because much of the FFAs being liberated during ischemic brain injury is believed to originate from phospholipids through the action of these enzymes, the Ca^{++} antagonists may be of benefit in attenuating FFA liberation and thereby, the severity of ischemic brain injury. Therefore, we have studied the effects of several calcium antagonists on the time course and magnitude of whole brain FFA liberation during decapitation ischemia in rats and compared them with the previously studied effects of pentobarbital.

Methods

Female Sprague-Dawley albino rats weighing 300 - 400 g were treated with (1) 4 ml 0.9% NaCl; (2) D-600, 20 mg/kg, IP; (3) lidoflazine 10 mg/kg, PO; (4) cinnarizine 10 mg/kg, PO and (5) nifidipine 20 mg/kg, PO and decapitated 2 h later. Immediately after decapitation, the heads were sealed in plastic bags containing saline-soaked sponges at $37^{o}C$ and placed into the incubator for 4 to 60 min or the brains removed from the calvaria and dropped into liquid N_2 at 30 s post-decapitation. The brains kept in the incubator were frozen in liquid N_2 at 4 to 60 min post-decapitation.

In another series of studies, rats were treated with lidoflazine, cinnarizine and nifidipine at various doses. Two hours later they were decapitated and their brains kept at $37^{o}C$ for 10 min post-decapitation before freezing in liquid N_2. The doses of various drugs were as follows:

lidoflazine and cinnarizine: 5, 10, 20, and 30 mg/kg;
nifidipine: 2.5, 5, 10, and 20 mg/kg.

Whole brain FFAs were quantitated as previously described by the methylesterification and GLC.

Results

The attenuation of whole brain FFAs by pentobarbital anesthesia has been previously published (5) but is presented for comparison with the effects of the calcium antagonists (Table 1). Pentobarbital anesthesia significantly attenuated total FFA liberation at all

Table 1 Effect of pentobarbital and nifidipine on whole brain free fatty acid accumulation after decapitation ischemia in rats. Values are $\overline{X}$ ± SEM (n = 4-8)

Drug	Ischemia Time (min)	16:0	18:0	18:1	20:4	Total
Normal Saline	0.5	55 ± 10.3	98 ± 8.1	57 ± 9.7	43 ± 7.5	253 ± 32.7
4 mg/kg	4	94 ± 12.8	196 ± 6.8	85 ± 7.4	156 ± 8.2	534 ± 32.2
I.P.	10	185 ± 4.7	306 ± 5.8	139 ± 5.0	260 ± 6.4	899 ± 19.0
	30	335 ± 17.5	538 ± 12.1	279 ± 10.2	403 ± 5.7	1578 ± 44.3
	60	508 ± 29.3	756 ± 21.1	423 ± 16.0	541 ± 18.3	2255 ± 80.2
Pentobarbital	0.5	41 ± 6.9	62 ± 4.9*	38 ± 3.5	21 ± 3.2*	167 ± 15.3*
60 mg/kg	4	82 ± 6.9	172 ± 4.5*	65 ± 5.0*	137 ± 4.7	461 ± 19.2
I.P.	10	154 ± 13.1*	260 ± 10.7*	110 ± 7.0*	215 ± 8.6*	745 ± 35.4*
	30	284 ± 16.1	470 ± 15.4	227 ± 10.0*	353 ± 13.3*	1353 ± 51.4*
	60	374 ± 11.4*	629 ± 12.9*	325 ± 14.8	455 ± 8.5*	1794 ± 40.0*
Nifidipine	0.5	22 ± 2.2*	54 ± 4.8*	18 ± 1.4*	24 ± 5.9	117 ± 12.7*
20 mg/kg	4	66 ± 6.3	166 ± 2.9*	64 ± 4.6	142 ± 4.0*	437 ± 4.8*
P.O.	10	116 ± 11.8*	260 ± 4.5*	108 ± 3.8*	199 ± 12.8*	685 ± 27.7*
	30	261 ± 44.2	461 ± 26.5*	210 ± 24.5*	355 ± 17.0*	1288 ±107.9*
	60	348 ± 51.0*	655 ± 21.9*	336 ± 18.5*	490 ± 18.4	1837 ±105.7*

*$P < 0.05$ compared to 0.9% NaCl injected controls

durations of ischemia except after 5 min. The longer the duration of ischemia, the greater the degree of attenuation by pentobarbital. It had similar effects on each of the FFAs studied. The rate of rise in arachidonic and stearic acid in the first 19 min of ischemia was greater than that observed for oleic and palmitic acids. However, the effects of pentobarbital were essentially the same on all 4 FFAs except for palmitic acid which appeared to be less.

Of the five calcium antagonists studied, the effects of nifidipine on FFA accumulation during ischemia were more pronounced and similar to the effects of pentobarbital (Table 1). Indeed, the effects of nifidipine appeared to be more pronounced than those of pentobarbital. Notably, nifidipine significantly attenuated total FFA liberation after all durations of ischemia and the degree of attenuation was greater the longer the duration of ischemia as was observed for pentobarbital. Unlike the effects of pentobarbital, however, the degree of attenuation of arachidonic acid liberation was not as impressive while the attenuations of stearic, oleic and palmitic acids were comparable to the effects of pentobarbital.

The effectiveness of the other Ca^{++} antagonists, namely, lidoflazine, D-600, cinnarizine and flunarizine in attenuating FFA liberation during complete global brain ischemia were clearly not as great as those of both pentobarbital and nifidipine (Table 2). Aside from the relative ineffectiveness in attenuating FFA liberation, the most notable finding is that none of these Ca^{++} antagonists significantly attenuated FFA accumulation after 60 min of decapitation ischemia, which was most noticeable for pentobarbital and nifidipine.

Except for arachidonic acid, cinnarizine did not attenuate FFA accumulation beyond 10 min of ischemia. The degree of attenuation at the various times of ischemia was not impressive and after 60 min whole brain FFAs were almost identical to that observed in the controls. Essentially the same effects were observed with lidoflazine except that total FFA, stearic and oleic acids were significantly attenuated at 30 min of ischemia. D-600 had little effect on total FFA, arachidonic, palmitic, and oleic acid accumulation where stearic acid liberation was significantly attenuated after 4, 10 and 30 min of ischemia. The effects of flunarizine were similar to those observed for lidoflazine and cinnarizine with minimal effects on total FFA, oleic, arachidonic and palmitic acid accumulation. However, as with D-600 and cinnarizine, the attenuation of stearic acid was more significant than the other FFAs.

Comparison of the effects of saline injected controls with the glycerin: ethanol H_2O (8:1:1 by volume) vehicle for the Ca^{++} antagonists after 10 min of decapitation significant effects of the latter on palmitic, oleic, arachidonic and total FFA levels (Table 3).

The dose related to attenuation of FFA liberation was tested for nifidipine, cinnarizine, and lidoflazine after 10 min of decapitation ischemia, and their effects were compared to the

Table 2 Effect of calcium antagonists on whole brain free fatty acid liberation during decapitation ischemia in rats. Values are $\bar{X}$ ± (SEM) (n = 4). *P < 0.05 compared to 0.9% NaCl injected controls

Drug	Ischemia Time (min)	16:0	18:0	18:1	20:4	Total
Cinnarizine	0.5	33 ± 3.9	61 ± 3.7*	26 ± 2.3*	22 ± 3.1	144 ± 10.5*
10 mg/kg	4	70 ± 7.1	171 ± 4.9*	64 ± 4.3	139 ± 13.4	445 ± 28.3
P.O.	10	142 ± 15.5*	257 ± 1.5*	113 ± 4.3*	201 ± 10.0*	717 ± 29.1*
	30	302 ± 26.7	499 ± 12.7	252 ± 8.6	374 ± 9.8*	1438 ± 51.5
	60	472 ± 16.8	735 ± 4.8	410 ± 12.3	532 ± 8.7	2163 ± 33.8
Lidoflazine	0.5	24 ± 3.1*	65 ± 1.9*	23 ± 1.1*	37 ± 2.9	149 ± 3.1*
10 mg/kg	4	120 ± 30.1	226 ± 39.2	105 ± 33.8	121 ± 19.2	575 ± 83.7
P.O.	10	137 ± 7.5*	285 ± 11.6	118 ± 6.0*	222 ± 11.7*	765 ± 31.5*
	30	298 ± 6.0	496 ± 9.1*	244 ± 4.7*	386 ± 12.0	1427 ± 13.7*
	60	450 ± 11.8	718 ± 8.5	388 ± 14.8	553 ± 27.8	2121 ± 57.9
D-600	0.5	38 ± 6.4	89 ± 12.1	44 ± 7.2	39 ± 13.4	213 ± 33.0
20 mg/kg	4	96 ± 12.8	175 ± 6.5*	89 ± 11.5	143 ± 4.3	513 ± 42.0
I.P.	10	149 ± 16.4*	271 ± 7.6*	126 ± 6.9	232 ± 6.0*	787 ± 33.2*
	30	304 ± 21.3	485 ± 10.1*	247 ± 12.3	370 ± 15.2	1423 ± 55.7
	60	398 ± 55.8	700 ± 41.6	379 ± 24.0	516 ± 28.7	2021 ± 143
Flunarizine	0.5	41 ± 6.5	76 ± 4.2*	37 ± 5.2	41 ± 4.3	196 ± 5.7
20 mg/kg	4	79 ± 7.3	175 ± 6.3*	69 ± 3.4	133 ± 6.9	460 ± 16.0
P.O.	10	166 ± 9.1	282 ± 9.6*	127 ± 8.4	225 ± 9.7*	806 ± 33.4*
	30	284 ± 13.2*	478 ± 7.0*	251 ± 7.1	339 ± 9.3*	1370 ± 17.3*
	60	441 ± 31.7	705 ± 16.2	390 ± 11.5	505 ± 14.1	2061 ± 64.7

Table 3   Whole rat brain FFA after 10 min decapitation ischemia in control

rats

	16:0	18:0	18:1	20:4	Total
0.9% NaCl					
4 mls/kg, IP	185 ± 5	306 ± 6	139 ± 6	260 ± 6	899 ± 19
(n = 8)					
Ca^{++} antagonist vehicle[+] 6 mls/kg, PO	153 ± 8[*]	296 ± 7	122 ± 3[*]	232 ± 8[*]	803 ± 19[*]
(n = 7)					

[+](glycerin:ethanol:H_2O (8:1:1 by vol) used as vehicle for Ca^{++} antagonists administered PO

[*]$P < 0.05$ compared to value in rats treated with 0.9% NaCl, IP

glycerin ethanol injected controls (Table 4). Nifidipine significantly reduced total FFA at a dose of 20 mg/kg, PO, but not at lower doses for the individual FFAs quantitated. However, oleic acid liberation was significantly reduced at doses of 5, 10, and 20 mg/kg. Lidoflazine was noticeably devoid of any dose-related effects on FFA liberation up to a dose of 30 mg/kg, PO. Cinnarizine showed a biphasic effect with greatest attenuation of FFA liberation at doses of 10 to 20 mg/kg without a significant effect at a dose of 30 mg/kg.

Discussion

Our findings show that the Ca^{++} antagonists tested have different effects on whole brain FFA liberation during complete global ischemia which also differs from the effects of pentobarbital. Compared to pentobarbital, except for nifidipine, most of the Ca^{++} antagonists were clearly less effective in attenuating FFA accumulation during ischemia suggesting that they may also be less effective as therapeutic agents for cerebral resuscitation although as we will discuss later, this is not necessarily the case.

Two aspects of the relative effects of pentobarbital and the Ca^{++} antagonists on FFA accumulation are of special interest. First, pentobarbital progressively attenuated FFA accumulation to a greater degree as the duration of ischemia was prolonged to 1 h. Although nifidipine had a similar effect, the other Ca^{++} antagonists did not show a similar effect. Instead, the degree of attenuation was only significant for durations of ischemia less than 30 min. This observation suggests that the Ca^{++} antagonists, namely, lidoflazine, flunarizine,

Table 4   Dose-related effects of calcium antagonists on whole rat brain free fatty acid liberation during 10 min decapitation ischemia.   Values are $\overline{X}$ ± SEM. (n = 4)

Drug	Dose	16:0	18:0	18:1	20:4	Total
Nifidipine PO	2.5 mg/kg	145 ± 14.7	279 ± 7.4	113 ± 7.9	236 ± 10.0	782 ± 19.9
	5.0 mg/kg	139 ± 21.3	282 ± 10.9	111 ± 3.9*	235 ± 9.5	771 ± 42.2
	10 mg/kg	121 ± 14.9	273 ± 10.6	107 ± 6.2*	214 ± 15.3	715 ± 39.2
	20 mg/kg	116 ± 11.8*	260 ± 4.5*	108 ± 3.8*	199 ± 12.8*	685 ± 27.7*
Cinnarizine PO	5 mg/kg	142 ± 15.8	272 ± 11.0	125 ± 13.8	209 ± 5.1	753 ± 40.7
	10 mg/kg	142 ± 15.5	257 ± 1.5*	113 ± 4.3	201 ± 10.0*	717 ± 29.1*
	20 mg/kg	127 ± 16.4	268 ± 4.4*	108 ± 1.0*	210 ± 8.7	718 ± 22.4*
	30 mg/kg	143 ± 18.4	286 ± 9.9	105 ± 5.4*	242 ± 6.4	785 ± 32.5
Lidoflazine PO	5 mg/kg	145 ± 8.1	283 ± 13.0	111 ± 10.6	233 ± 8.7	778 ± 31.2
	10 mg/kg	137 ± 7.5	285 ± 11.6	118 ± 6.0	222 ± 11.7	765 ± 31.5
	20 mg/kg	146 ± 13.6	294 ± 6.4	140 ± 15.7	222 ± 9.6	807 ± 36.4
	30 mg/kg	140 ± 5.4	276 ± 9.8	112 ± 5.5	228 ± 9.3	760 ± 21.7

*$P < 0.05$ compared to 0.9% NaCl injected controls

cinnarazine, and D-600 affect one phase of FFA accumulation, namely, those mechanisms responsible for FFA liberation in the first 30 min of ischemia but not those liberating FFA after longer durations of ischemia. These differences suggest that the liberation of FFA during ischemia occurs through different mechanisms as the duration of ischemia progresses. Whereas pentobarbital and nifidipine attenuates the processes liberating FFA in both the early and late phases, the other Ca^{++} antagonists only affect the early phase.

Second, the differences in the time course of the accumulation of arachidonic and stearic acids as opposed to that of oleic and palmitic acids also suggest different mechanisms or origins of FFA release during ischemia. Both arachidonic and stearic acids rise rapidly in the first 10 min of ischemia and thereafter, at a slower but more linear rate for up to 60 min of ischemia. Oleic and palmitic acids, in contrast, rise almost linearly from the onset of ischemia and continue at the same rate as ischemia is prolonged. Thus, the rate of rise in arachidonic and stearic acids differs most markedly from that of oleic and palmitic acids in the first 10 min of ischemia. We speculate perhaps the early rise in arachidonic and stearic acids is related to neurotransmitter depolarization induced release from membrane phospholipids whereas in the later phase their origin, source, or mechanism of release changes to that from which oleic and palmitic acids are derived. Since at this time the rate of rise in arachidonic and stearic acids parallel the increase in these two FFAs. Perhaps in the early phase both arachidonic and stearic acids arise from membrane phospholipids via the action of glyceride lipases, which may also be the origin of oleic and palmitic acids at the outset. However, precise localization of the origins and mechanisms of FFA release remain to be defined.

The limitation of our model as previously mentioned, specially regarding the efficacy of the Ca^{++} antagonists is that if they are of benefit primarily through their effects on improving perfusion through vascular effects, it would not be revealed in this model.

References

1. Bleyaert AL, Nemoto EM, Safar P, Stezoski SW, Mickell JJ, Moossy J, Rao GR (1978) Thiopental amelioration of brain damage after global ischemia in monkeys. Anesthesiology 49:390-398

2. Goldstein A, Wells BA, Keats AS (1966) Increased tolerance to cerebral anoxia by pentobarbital. Arch Int Pharmacodyn 161:138-143

3. Nemoto EM, Shiu GK, Nemmer JP, Bleyaert AL (1982) Attenuation of brain free fatty acid liberation during global ischemia: A model for screening potential therapies for efficacy? J Cereb Blood Flow Metabol 2:475-480

4. Shiu GK, Nemoto EM (1981) Barbiturate attenuation of brain free fatty acid liberation during global ischemia. J Neurochem 37:1448-1456

54

5.	Shiu GK, Nemmer JP, Nemoto EM (1982) Reassessment of brain free fatty acid liberation during global ischemia and its attenuation by barbiturate anesthesia. J Neurochem (submitted for publication)

6.	Shiu GK, Nemoto EM, Nemmer JP (1982) Dose-related attenuation of brain free fatty acid liberation during global ischemia by various anesthetics. Crit Care Med (submitted for publication)

Alterations in Whole Brain Cyclic-AMP and Cerebral Cortex Na-Inducible Cyclic-AMP in Rats During and After Complete Global Ischemia

M. R. Lin, E. M. Nemoto, and P. D. Kessler

The Anesthesia and CCM Research Laboratories, Department of Anesthesiology and
Critical Care Medicine, University of Pittsburgh, School of Medicine,
1081 Scaife Hall, Pittsburgh, PA 15261, USA

Introduction

Neurologic dysfunction after cerebral ischemic anoxic insults may be due not only to
irreversible neuronal necrosis and infarction but also, a possibly reversible failure in
synaptic transmission. The EEG remains abnormal despite complete recovery of brain high
energy phosphates after recirculation following 30 min of complete global ischemia in dogs
(5). Persistent EEG alterations correlate with persistent alterations in brain catecholamine
metabolism despite recovery of brain high energy phosphates (2). These findings show that
there may be a disassociation between recovery of brain oxidative metabolism, reflected by
brain high energy phosphates and synaptic transmission, reflected by the EEG.

Failure in brain energy metabolism during cerebral ischemic anoxic insults is believed to
cause the release of cerebral neurotransmitters (NT) in massive amounts with three primary
detrimental effects. First, although the NT may be of no consequence in metabolically
inactive tissue such as during complete global ischemia, with reoxygenation it could result in
a hypermetabolic state that exceeds the ability of the circulation to meet its metabolic
demands. Second, the NT may lead to inappropriate cerebrovascular responses such as
vasospasm. Third, the exposure of NT receptors to exceedingly high concentrations of NT
may lead to their desensitization and attentuated response. All of these factors are believed
to be important impediments to neuronal recovery after cerebral insults. The combination of
postischemic hypermetabolism and delayed hypoperfusion are recognized as important
factors in complicating neuronal postischemic recovery (7,12).

Adenosine 3',5' monophosphate (cyclic-AMP) mediates many of the physiological and
biochemical processes in synaptic transmission (3). Thus, brain cyclic-AMP accumulation
directly correlates with pharmacological, electrical or injury-induced cerebral activation
(9,11). One of the aims of this study was to assess the time course and magnitude of changes
in whole brain cyclic-AMP during and at various times after complete global brain ischemia
to gain insight into the patterns of cerebral metabolic activity with recirculation following
complete global brain ischemia. The effects of thiopental loading following complete global
ischemia on brain cyclic-AMP were also evaluated.

Changes in receptor sensitivity after recirculation could affect the efficiency of synaptic transmission and result in neurologic dysfunction. It should be emphasized, however, that a failure in synaptic transmission could result not only from alterations in receptor sensitivity but also defects in NT synthesis and metabolism or release and reuptake mechanisms as a direct result of the cerebral insult. Our aim in the second series of studies was to assess the responsivity and viability of mechanisms involved in synaptic transmission on the basis of in vitro testing of NA inducible cyclic-AMP accumulation in rat frontal-parietal cerebral cortex at various times during and after complete global brain ischemia.

Methods

Female S-D albino rats weighing 300 to 400 grams with free access to food and water were used in all studies. In the first series, rats were subjected to global ischemia by either decapitation or a high pressure neck cuff (1500 torr) combined with arterial hypotension (MAP = 50 torr). Rats were decapitated unanesthetized but for the neck cuff, they were anesthetized with 1.0% halothane, 66% N_2O and 33% O_2. At various times during and after ischemia, the brains were rapidly sampled into liquid N_2 using a technique we developed. Whole brain cyclic-AMP was assayed by the protein binding method of Gilman. In a separate group of rats, the effects of thiopental loading (90 mg/kg, IV) on whole brain cyclic-AMP levels after recirculation was studied.

In a second series of the study, unanesthetized rats were decapitated or rats anesthetized with 66% $N_2O/33\%$ O_2 were subjected to various durations of ischemia and periods of recirculation. Thereafter, the magnitude of cyclic-AMP accumulation to NA stimulation in the cerebral cortex slices was tested in vitro as described by Baudry et al. (1). The methods for the preparation of the slices, and incubation conditions have been previously described (6). Cyclic-AMP, protein, ATP and creatine phosphate (CP) were assayed.

Results

In anesthetized rats, whole brain cyclic-AMP was 1.17 + 0.05 (SEM) nmoles/g brain which compares favorably with values measured in brains inactivated by freeze-blowing and focussed microwave irradiation (4).

During decapitation ischemia, whole brain cyclic-AMP increased 3-fold in 2 min and fell below normal after 8 min with little change thereafter (Fig. 1). Cyclic-AMP also increased by about 3-fold in rats subjected to tourniquet ischemia during halothane anesthesia. However, the peak level occurred after 4 instead of 2 min of ischemia. By 16 min, cyclic-AMP was below preischemic levels.

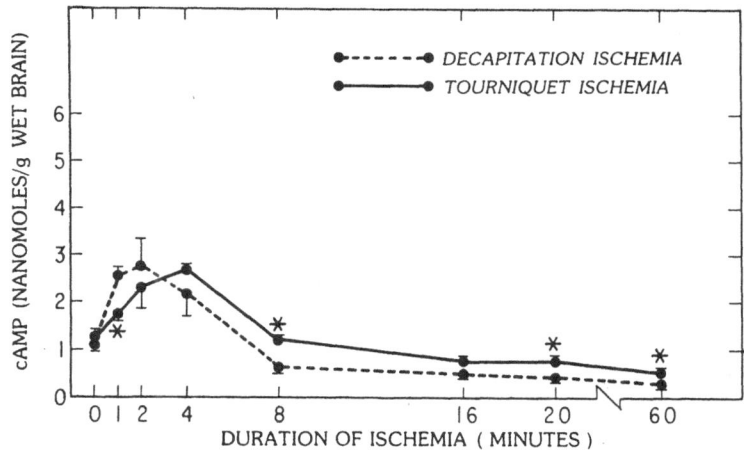

Fig. 1   Whole rat brain cyclic-AMP during complete global brain ischemia by decapitation and neck tourniquet. Rats were decapitated unanesthetized. Tourniquet
ischemia was induced in anesthetized rats mechanically ventilated on 1%
halothane and oxygen. Ischemia was induced by trimethaphan-induced arterial
hypotension to a mean arterial pressure of about 50 torr plus inflation of a high
pressure (1500 torr) neck tourniquet. * = P≤ 0.05 compared to corresponding
values in decapitated rats. ⊥ = SEM. Decapitation ischemia: n = 6 for each point
except at 1 and 16 min where n = 3 and 1 respectively. Tourniquet ischemia: n =
6 to 12 for each point except at 4, 16 and 60 min where n = 3, 2 and 4
respectively

In rats subjected to tourniquet ischemia with recirculation, peak cyclic-AMP levels occurred
after 5 min of recirculation (Fig. 2). Following 8 min of ischemia, it rose by 5-fold and after
16 min of ischemia, by 13-fold. After 20 min of ischemia, the increase was attenuated and
only about 3-fold. Thus, the greatest rise in brain cyclic-AMP occurred after 16 min of
ischemia. The elevation of cyclic-AMP was also sustained longer during recirculation
following 16 min of ischemia.

Thiopental loading after 16 min of ischemia accelerated the fall in cyclic-AMP during
recirculation (Fig. 3). In untreated rats, brain cyclic-AMP was elevated by 6-fold after 30
min of recirculation and 4-fold after 60 min. In thiopental rats, it was back to normal after
30 min of recirculation. Thiopental loading significantly lowered brain cyclic-AMP compared
to controls at 30, 60 and 120 min postischemia.

Following 16 min of ischemia in untreated rats, brain lactate rose to about 17 μmoles/g
brain after 5 min of recirculation and remained elevated until 30 min postischemia (Fig. 4).
Thiopental significantly reduced brain lactate at 15 and 30 min postischemia compared to
controls but levels were similar in both groups at 60 and 120 min.

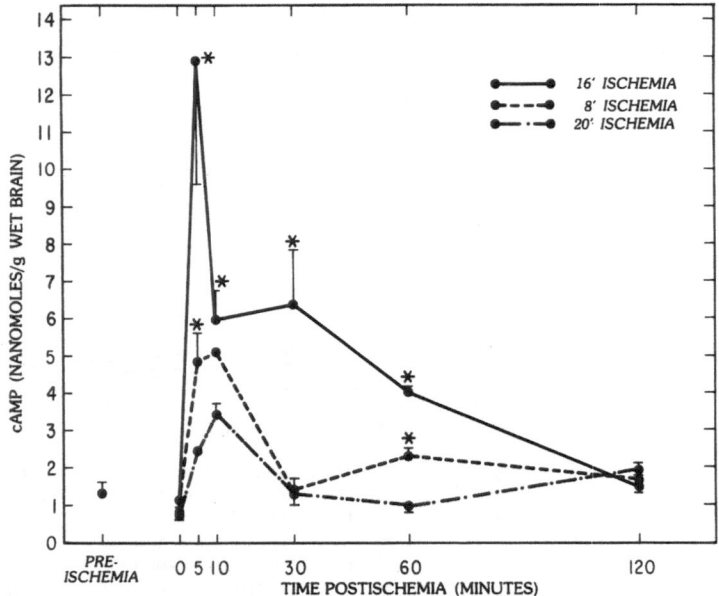

Fig. 2   Postischemic changes in whole rat brain cyclic-AMP after various durations of complete, transient global brain ischemia. Global ischemia was induced in halothane anesthetized rats by a combination of arterial hypotension and a high pressure neck tourniquet. Anesthesia was discontinued during ischemia and the rats ventilated on 100% oxygen. * = P ≤ 0.05 compared to preischemic values. ⊥ = SEM. Two to 4 rats at each point except for one rat each at 5 min following 20 min of ischemia and at 10 min following 8 min of ischemia. Preischemic value is the mean of 7 rats

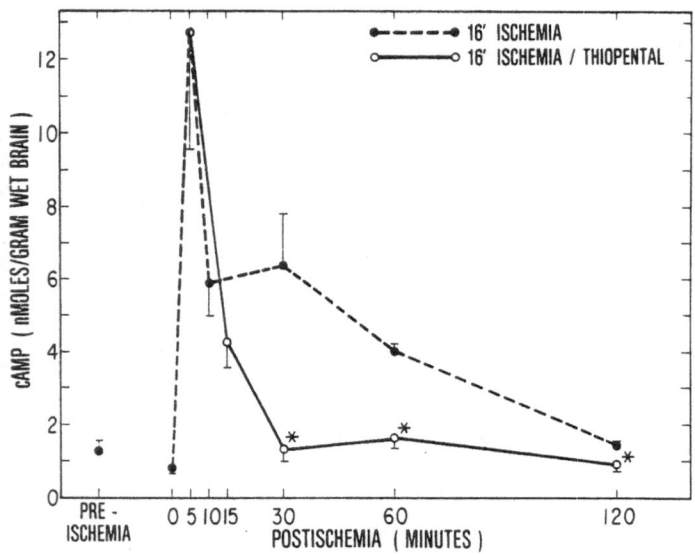

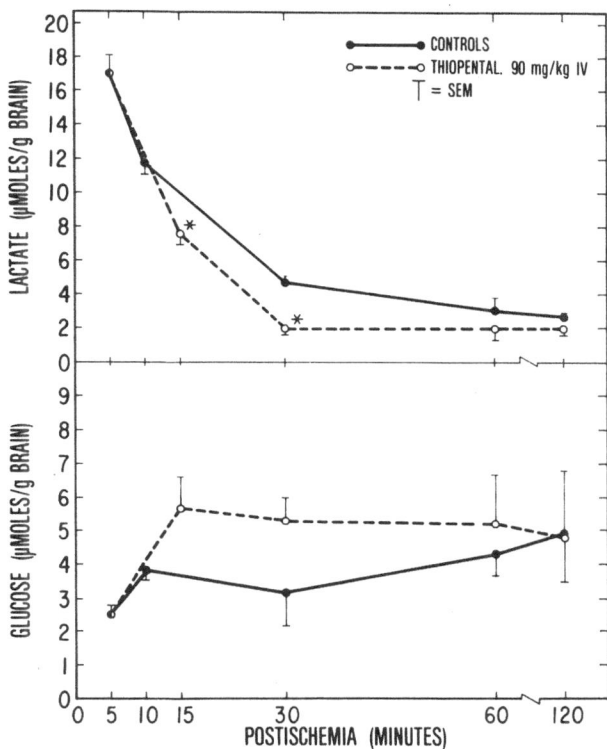

Fig. 4   Whole brain lactate and glucose after 16 min complete global brain ischemia. See legend Fig. 2 for other details. ∗ = P ⩽ 0.05 compared to the corresponding values in untreated controls

◄ Fig. 3   Whole brain cyclic-AMP after 16 min complete global brain ischemia induced by a combination of arterial hypotension and a high pressure neck cuff. Thiopental 90 mg/kg was infused IV beginning at 5 min after recirculation. One third the total dose was infused in 5 min and two thirds- over the ensuing 55 min. Preischemia, n = 7. ∗ = P ⩽ 0.05 compared to values in the control untreated group

60

In untreated rats, brain glucose rose to about 4 µmoles/g brain between 10 and 120 min postischemia (Fig. 4). In thiopental rats, it rose to between 5 and 6 µmoles/g brain after 60 min of recirculation but the difference from the controls progressively diminished as the duration of recirculation was prolonged.

In the second series of the study, we found that cyclic-AMP increased by about 10-fold between 0 and 20 min of ischemia in unstimulated cortical slices (Fig. 5). It was unchanged between 30 and 45 min of ischemia, then increased about 25-fold to 750 pmoles/mg protein between 45 and 60 min. Cyclic-AMP changes in NA-stimulated slices paralleled those in unstimulated slices.

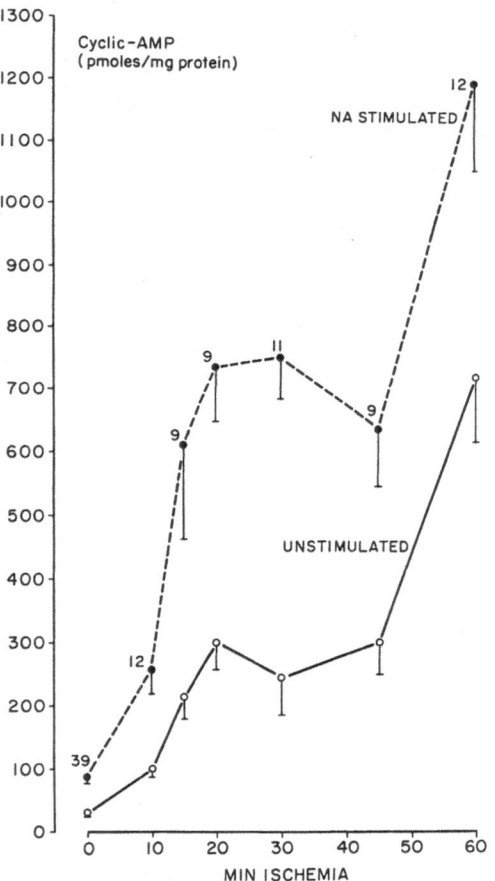

Fig. 5   Rat frontal-parietal cortex cyclic-AMP after various durations of ischemia with and without 11.2 micromolar noradrenalin stimulation in vitro. Numbers indicate the number of rats used for each point, in both stimulated and unstimulated slices. Statistical flags indicate SEM

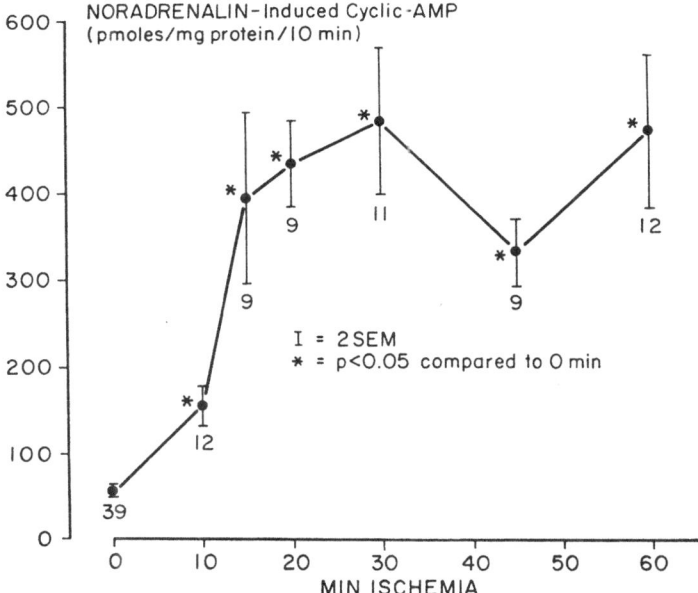

Fig. 6   Changes in noradrenalin (NA) inducible cyclic-AMP in rat cortex slices in vitro after various durations of decapitation global brain ischemia

NA-inducible cyclic-AMP, or the difference between NA-stimulated and unstimulated cyclic-AMP, rose by 8-fold between 0 and 20 min of ischemia and was essentially unchanged thereafter (Fig. 6).

Rats maintained on 66% N_2O/33% O_2 anesthesia with pancuronium immoblization and mechanical ventilation for up to 6 h showed a progressive rise in NA-inducible cyclic-AMP between 0 and 3 h without a further increase thereafter (Fig. 7). Fifteen min of ischemia caused an 8-fold rise in NA-inducible cyclic-AMP before recirculation. However, after 1 h of recirculation, it fell before beginning to rise to about 500 pmoles/mg protein/10 min. Twenty min of ischemia resulted in essentially the same pattern of changes except that the increase after 1 h of recirculation was not as great. Thirty min of ischemia, on the other hand, caused a progressive fall in NA-inducible cyclic-AMP with the start of recirculation until by 6 h, there was hardly any response at all.

After 6 h of recirculation and 0, 15, 20 and 60 min of ischemia, there was an increase in NA-inducible cyclic-AMP from about 300 pmoles/mg protein/10 min to about 600 pmoles/mg protein/10 min (Fig. 8). As ischemia was prolonged beyond 15 min, it declined with almost no response after 30 min of ischemia.

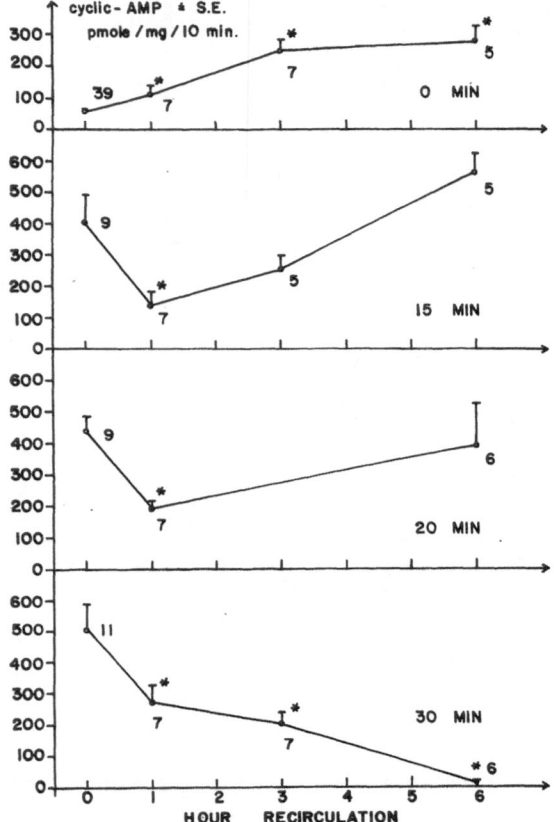

Fig. 7

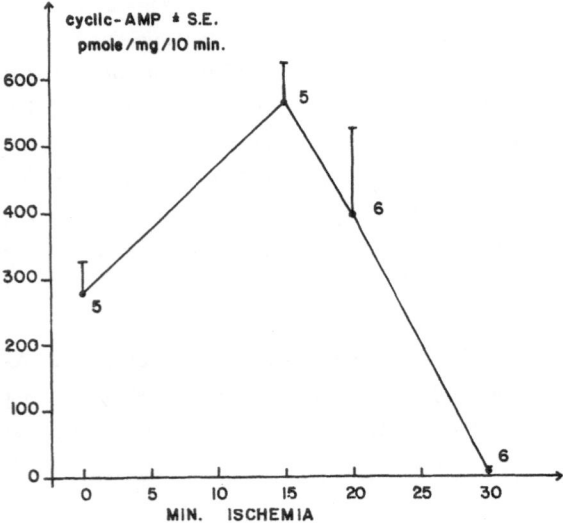

Fig. 8

◄ Fig. 7   Noradrenalin (NA) inducible cyclic-AMP in rat cerebral cortex slices after
various durations of ischemia and time of recirculation. Complete global brain
ischemia was induced by a combination of a high pressure neck tourniquet and
arterial hypotension. Numbers indicate the number of rats used for each point. ₊
= P ≤ 0.05 compared to value at 0 min of recirculation

Fig. 8   Noradrenalin (NA) inducible cyclic-AMP in rat cerebral cortex tested in vitro
after various durations of ischemia and 6 h of recirculation in vivo.. Numbers
indicate the number of rats studied for each point. Statistical flags indicate
SEM

The point of inflection with regard to the NA-inducible cyclic-AMP changes, namely at 15
min of ischemia corresponded with a cross-over point for levels of ATP and CP (Fig. 9).
Although the values of the high energy phosphates are only preliminary, they suggest that
beyond 15 min of ischemia, a critical change may be occurring.

Discussion

The cyclic-AMP response of the brain to acute injury appears to be a function of the
severity of the insult, the integrity of brain oxidative metabolism (8), and presumably, the
sensitivity of the receptor-adenylate cyclase system. Based on this assumption, our
interpretation of the finding that the greatest rise in brain cyclic-AMP occurred with
recirculation after 16 min of ischemia rather than 8 or 20 min is as follows. Eight min of
ischemia may not have produced an insult of maximum severity and therefore, of maximum
stimulation of adenylate cyclase. Thus, the rise in cyclic-AMP was less than that observed
after 16 min of ischemia. After 20 min of ischemia, the severity of the injury was probably
of sufficient magnitude or at least comparable to that induced by 16 min of ischemia but the
integrity of cerebral metabolism may have been compromised such that it was unable to
respond to the stimulus with accelerated generation of cyclic-AMP. Thus, these data would
suggest that between 16 and 20 min of ischemia, there is a critical threshold for a drastic
alteration in cerebral metabolism which may or may not be reversible after restoration of
circulation.

The fact that thiopental loading suppressed both the cyclic-AMP accumulation and the
lactate levels with improved brain glucose resaturation suggests several possible
mechanisms of action. First, that thiopental suppressed the hypermetabolic response after
recirculation as evidenced by its suppression of brain cyclic-AMP levels. Second, that
thiopental improved brain circulation at least in areas of impaired perfusion even after
perfusion pressure to the brain was restored thus resulting in acceleration of brain lactate
clearance and improved resaturation with glucose. Third, thiopental may have suppressed
the hypermetabolic response after recirculation resulting in attenuated lactate production
and conversion of glucose to lactate which would result in higher glucose levels. Because the

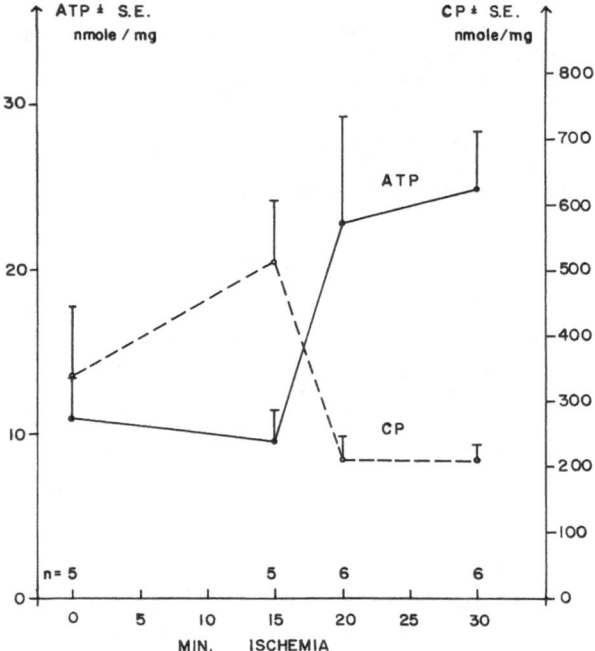

Fig. 9   Adenosine triphosphate (ATP) and creatine phosphate (CP) levels in cerebral cortex slices in vitro from rats subjected to various durations of complete global brain ischemia by a combination of a high pressure neck tourniquet and arterial hypotension during nitrous oxide, 66% anesthesia and 6 h of recirculation. Statistical flags indicate SEM

effects of thiopental were primarily demonstratable in the early period after recirculation it is likely that the primary effect of thiopental is attributable to its suppression of the hypermetabolic response upon recirculation.

Our findings of an 8 to 10-fold increase in NA-inducible cyclic-AMP during complete global brain ischemia of up to 60 min duration without recirculation were surprising. First, we had hypothesized that NT released during the ischemic insult would result in desensitization of receptors and we had expected to find an attentuation of the response during ischemia. Our data show that exactly the opposite is true. Namely, that there is a rapid increase in NA-inducible cyclic-AMP in the first 15 to 20 min of ischemia after which time there is neither a decrease nor increase with longer durations of ischemia. Schwartz et al. (10) showed that in gerbil brain homogenate, adenylate cyclase activity was unaltered during ischemia, but was reduced to 60% of normal during 20 h of recirculation. It returned to normal by seven days postinsult. With regard to the changes occurring during ischemia, the relevance of these changes in a membrane bound enzyme in a broken cell preparation to changes occurring in situ is questionable. The lack of a change in enzyme activity only indicates that there was no permanent alteration in enzyme function. However, the environment must play an important role in modulating enzyme activity in situ. Our results may also be explained by a defect in the reuptake of catecholamines into synaptic vesicles and an inhibition of monoamine oxidase both of which would result in higher concentrations of noradrenalin to which the tissue would be exposed when the NA is added in vitro. Again, as in the previous series of studies, our findings in this series suggest that somewhere between 15 and 20 min of ischemia a critical alteration occurs resulting in a maximal rise in NA-inducible cyclic-AMP response which is not further altered with up to 1 h ischemia.

Despite the sustained exaggerated response to NA during ischemia without recirculation of up to 1 h duration, this appears to be the case only if recirculation does not occur. With recirculation, the changes in NA-inducible cyclic-AMP accumulation were dependent upon and affected by both the time of recirculation and the duration of ischemia. As observed in our first series of studies, the peak response in NA-inducible cyclic-AMP occurred after 6 h of recirculation when the duration of ischemia was 15 min. Increasing the duration of ischemia to 20 min resulted in an attenuation of the response at 6 h of recirculation compared to that observed after 15 min of ischemia. After 6 h of recirculation following 30 min of ischemia, the response was almost totally obliterated. The mechanisms involved in this apparent desensitization to NA are unclear but again, these findings like the others also suggest that a critical metabolic alteration occurs between durations of ischemia of 15 and 20 min. In any case, we believe that our findings have provided evidence of a delayed alteration in cerebral viability after durations of ischemia which may be related to defects in the processes involved in synaptic transmission.

References

1. Baudry M, Martres MP, Schwartz JC (1976) Modulation in the sensitivity of noradrenergic receptors in the CNS studied by the responsiveness of the cyclic-AMP system. Brain Res 116:111-124

2. Brown RM, Carlson A, Ljunggren B, Siesjö BK, Snider SR (1974) Effect of ischemia on monoamine metabolism in the brain. Acta Physiol Scand 90:789-791

3. Greengard P (1978) Cyclic nucleotides, phosphorylated proteins and neuronal function. Raven Press, New York

4. Guidotti A, Cheney DL, Trabucchi M, Doteuchi M, Wang C (1974) Focussed microwave radiation: A technique to minimize post mortem changes of cyclic nucleotides, dopa and choline and to preserve brain morphology. Neuropharmacol 13:1115-1122

5. Hinzen DH, Müller U, Sobotka P, Gebert E, Lang R, Hirsch H (1972) Metabolism and function of dog's brain recovering from longtime ischemia. Am J Physiol. 223:1158-1164

6. Lin MR, Henteleff HB, Nemoto EM (1982) Noradrenalin-inducible cyclic-AMP accumulation in rat cerebral cortex: Changes during complete global ischemia. J Neurochem (submitted for publication)

7. Nemoto EM (1978) Pathogenesis of cerebral ischemia-anoxia. Crit Care Med J 6:203-214

8. Passonneau JV, Kobayashi M, Lust WD (1977) The effect of bilateral ischemia and recirculation on energy reserves and cyclic nucleotides in the cerebral cortex of gerbils. In: Thurman, Williamson, Drott, Chance. Eds. Alcohol and Aldehyde Metabolizing Systems. Vol. III. Academic Press, New York, pp 485-498

9. Schultz J, Daly JW (1973) Accumulation of cyclic adenosine 3',5' monophosphate in cerebral cortical slices from rat and mouse: Stimulatory effect of alpha and beta adrenergic agents and adenosine. J Neurochem 21:1319-1326

10. Schwartz JP, Mrsulja BB, Mrsulja BJ, Passonneau JV, Klatzo I (1976) Alterations of cyclic nucleotide-related enzymes and ATPase during unilateral ischemia and recirculation in gerbil cerebral cortex. J Neurochem 27:101-107

11. Shimizu H, Takenoshita M, Huang M, Daly JW (1973) Accumulation of adenosine 3',5'-monophosphate in brain slices: Interaction of local anesthetics and depolarizing agents. J Neurochem 20:91-95

12. Snyer JV, Nemoto EM, Carroll RG, Safar P (1975) Global ischemia in dogs: Intracranial pressures, brain blood flow and metabolism. Stroke 6:21-27

Pathophysiology and Pathobiochemistry of Acute Brain Infarction in the Gerbil: The Influence of Metabolic Inhibition*

G. Mies, H.-J. Bosma, W. Paschen, and K. A. Hossmann

Max-Planck-Institut für Neurologische Forschung, Abteilung für Experimentelle Neurologie, 5000 Köln 91, FRG

Introduction

The understanding of the pathophysiology of stroke requires the use of experimental models which under standardized conditions are able to mimic the clinical situation as closely as possible. Among the various models used, the gerbil (Meriones unguiculatus) is of particular interest because in this species the circle of Willis is incomplete, and extracranial carotid artery occlusion results in stroke in a certain percentage of animals (11). A certain disadvantage, however, is the fact that size and location of the infarct vary considerably. Therefore, any generalization is precluded unless statistical evaluation of large series of animals is carried out.

For this reason, methods in which various hemodynamic and metabolic parameters can be evaluated simultaneously in the same animal have found increasing interest. This is because these methods allow a valid correlation of the factors involved in individual experiments. An example is the double-labeled autoradiography of blood flow and glucose utilization (15), which can be combined with bioluminescent and fluoroscopic techniques for regional assessment of glucose, ATP and NADH distribution on intact brain sections. In the first part of this investigation, this approach was used for establishing the hemodynamic and pathobiochemical correlates of the neurological deficits observed.

Based on these findings, the influence of therapeutic interference was studied in the second part of this report. Several authors have described an ameliorating effect of metabolic inhibition by barbiturates or hypothermia, induced after the onset of focal ischemia (1,8,9,13,14,20,28). The pathobiochemical correlates of the neurological status are known. We were interested to find out if and under what conditions this approach improves the sequela of induced ischemia in the present experimental situation. The results obtained indicate that metabolic inhibition is of very limited use for improving the final outcome of experimental stroke if not applied within a well-defined interval after the onset of ischemia.

*   This study was partly supported by the Deutsche Forschungsgemeinschaft Pa 266/2-2

Materials and methods

Regional evaluation of blood flow and metabolism

In fifteen gerbils (Meriones unguiculatus) of both sexes weighing 60-80 g, the right common carotid artery and the left external carotid artery were occluded permanently under anesthesia with 1.2% halothane delivered by a gas mixture of 30% oxygen and 70% nitrogen. After surgery, animals were allowed to recover and were classified as described previously (3): asymptomatic animals did not suffer from any visible neurological deficits; animals with mild symptoms exhibited unilateral hemiparesis, and animals with severe symptoms, suffered rolling seizures in addition.

Sixty minutes after onset of vascular occlusion, the gerbils were again anesthetized, and femoral arteries and veins were catheterized for infusion of tracers, for monitoring of arterial blood pressure, and for arterial blood sampling. Body temperature was kept constant at $37^{\circ}C$ by controlled heating with an infrared light source.

After the surgical procedure, a bolus of 15 µCi ^{14}-deoxyglucose dissolved in 0.3 ml Ringer solution was given intravenously for measurement of glucose utilization according to Sokoloff et al. (23). Thirty minutes later, cerebral blood flow was determined with the iodo-antipyrine technique described by Sakurada et al. (18). 100 µCi ^{131}I-iodo-antipyrine dissolved in 0.5 ml Ringer solution was infused intravenously over a period of one minute, at the end of which the animals were immersed in liquid nitrogen. Brains were removed in a cold box at $-20^{\circ}C$ and 20 µm thick sections were produced in a cryostat at the same temperature for autoradiography. Alternate brain sections were freeze-dried and processed for evaluation of regional glucose and ATP concentration by means of bioluminescent methods (10,16). Finally, the tissue block in the cryostat was illuminated with ultraviolet light for visualization of NADH according to Welsh and Rieder (27).

By choosing ^{131}I and ^{14}C as the radioactive labels, it is possible to perform double-tracer autoradiography from the same brain section. At high concentration of ^{131}-I-iodo-antipyrine contamination by ^{14}C can be neglected and exposure of the brain sections results in autoradiograms which represent cerebral blood flow. After decay of ^{131}I (within 8 weeks), brain sections can be exposed to ^{14}C radioactivity for evaluation of glucose consumption (15).

Bioluminescent techniques are based upon substrate specific reactions which result in light emission. The freeze-dried sections are covered with a slice of frozen enzyme solution. When this sandwich is warmed up, the enzyme solution penetrates into the section, and the resulting bioluminescent reaction can be recorded on photographic film. Finally, NADH fluoroscopy is used as an indicator of regional redox state of the brain.

Treatment of focal ischemia with barbiturates or hypothermia

Twenty gerbils did not receive treatment after carotid occlus.on and served as ischemic controls. Three hundred gerbils were submitted to the following procedures after vascular occlusion:

a)   Pentobarbital was applied intraperitoneally at a dose cf 20, 30, 40 or 50 mg/kg immediately (n = 30 for each dose) or 5 min (n = 30 for each dose) after vascular occlusion. The body temperature was kept constant at $37^{o}C$ by means of controlled heating.

b)   Hypothermia was induced 5 min after vascular occlusion by immersing the anesthetized animal in cold water until body temperature reached $28^{o}C$. Hypothermia was maintained at this level with a feedback-controlled infrared light source for one, two or three hours respectively (n = 20 for each group).

Statistical evaluation of therapeutic interference was based cn the survival rate of the animals after ischemia. Treated and untreated groups were compared using the Fischer's exact F test.

Results

Regional findings in acute focal ischemia

In earlier studies, it has been demonstrated that after the combined occlusion of one common carotid artery and the opposite external carotid artery, 30% of the gerbils exhibit no neurological deficits. In 70% of the animals, however, reurological symptoms are detected allowing the classification into a group of gerbils with mild symptoms (35%) and into a group with severe symptoms (35%) (3).

Regional findings of blood flow and energy metabolism representative for the different groups of animals 2 hours after onset of focal ischemia are shown in Fig. 1. In an animal which did not develop any neurological symptoms, vascular occlusion led to a reduction of blood flow with a concomitant depletion of tissue glucose and ATP and a rise in NADH fluorescence restricted to the hippocampus. In other subcortical structures of the right hemisphere, blood flow and glucose content but not ATP were diminished. The increase of glucose consumption in this area indicates that the incipient oxygen deficiency was compensated by activated glycolysis. In an animal with mild symptoms, the whole right hemisphere exhibited a decrease of tissue glucose and ATP content and a rise in NADH fluorescence. Blood flow was drastically reduced in the cortex and subcortical structures. The increased phosphorylation of deoxyglucose most probably reflected anaerobic glycolysis. In an animal with severe symptoms, however, a complete breakdown of energy metabolism was observed in the right as well as in the marginal part of the left hemisphere. Glucose consumption in contrast to the other symptomatic group had ceased.

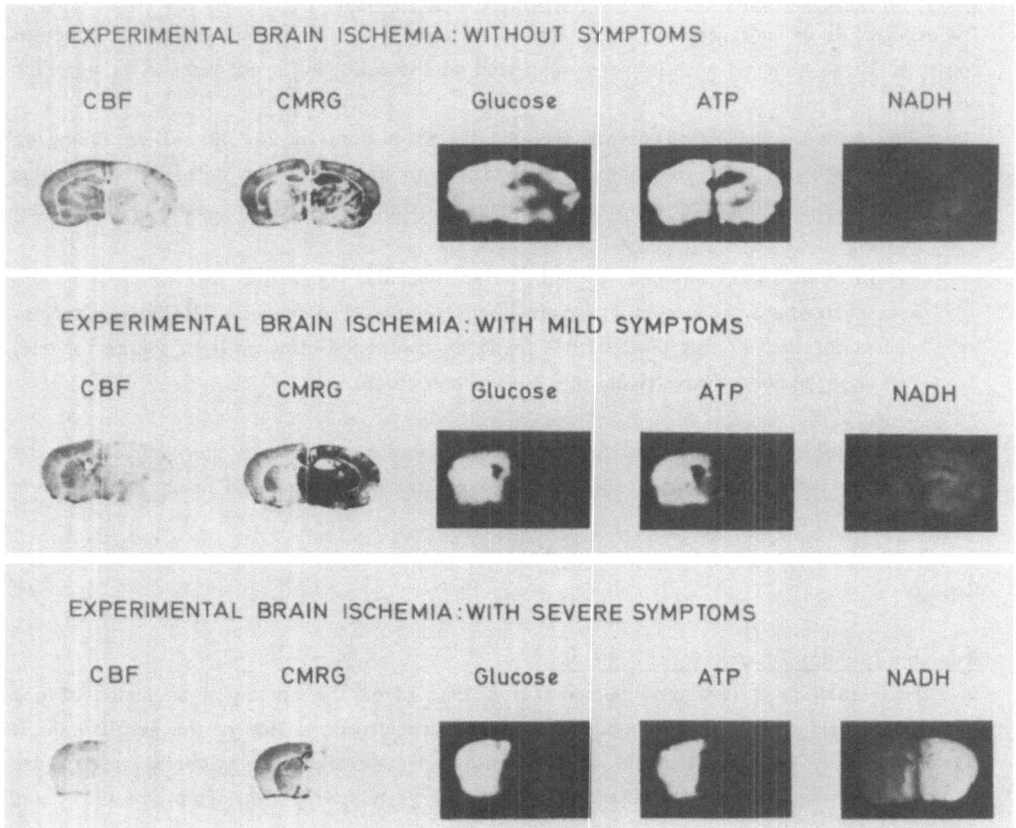

Fig. 1   Representative regional findings of cerebral blood flow, glucose utilization, glucose and ATP content and NADH concentration 2 hours after onset of focal ischemia in the gerbil. Note the circumscribed disturbance of blood flow and energy metabolism in animals without neurological symptoms; in symptom-positive gerbil, the right hemisphere is affected by the energy failure

Effect of pentobarbital and hypothermia on the survival rate

As demonstrated in Figure 2, the intraperitoneal barbiturate application of 50 mg/kg body weight given immediately after vascular occlusion was the only dose which significantly altered the survival rate during the observation time of 30 days. Lower doses of 30 and 40 mg/kg pentobarbital led only to a transient effect during the early phase of focal ischemia.

In another series of experiments, barbiturates were administered intraperitoneally 5 min after vascular occlusion. However, the delayed application of pentobarbital did not improve the survival rate significantly as compared to the untreated control group.

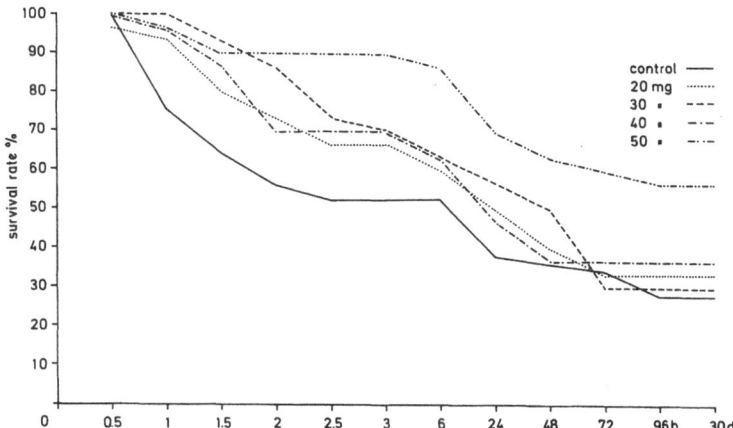

Fig. 2   Survival rate after permanent modified occlusion of the carotid arteries in the gerbils treated immediately with a single dose of pentobarbital. Note that only a dose of 50 mg/kg was effective to decrease mortality significantly ($p < 0.05$)

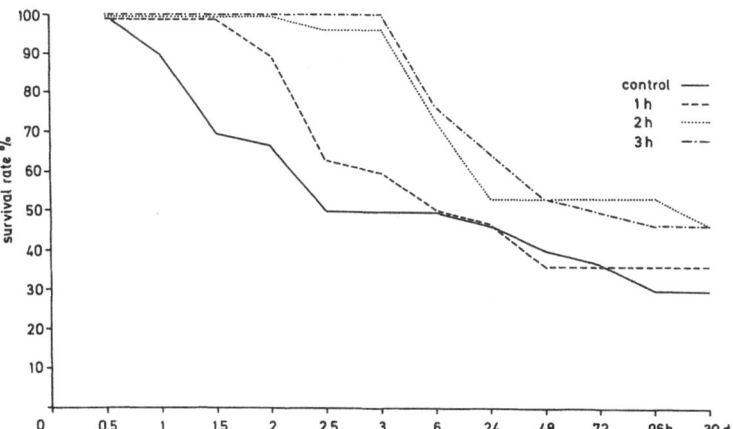

Fig. 3   Survival rate after permanent modified occlusion of the carotid arteries in the gerbil treated with hypothermia for 1, 2 and 3 hours. Note that the 100% survival is a function of the hypothermia duration which does not significantly improve the survival rate after 30 days

During hypothermia to 28°C lasting 1, 2 or 3 hours, none of the animals died whereas in the control group mortality was between 50 and 60%. When normothermia was reestablished, mortality of the gerbil rapidly increased again and there were no significant differences in the survival rate of treated and untreated animals after rewarming (Fig. 3).

Discussion

Ischemia of the brain is the result of a disproportion between the high cerebral energy demand and the critically reduced substrate availability due to low blood flow. The therapeutic concept to avoid progressing ischemic tissue damage, should therefore consider improvement of substrate supply by means of increasing blood flow on the one hand and/or reduction of energy metabolism by inhibition or hypothermia on the other hand. So far, therapeutic approaches for increasing blood flow to the ischemic area have not been very successful (2). Therefore, decreasing brain metabolism by hypothermia (17), or barbiturates (4,13) has been proposed to improve the survival of ischemic brain tissue.

Cerebral hypothermia is known to decrease brain metabolism directly depending on the temperature, thus inducing protection against pathological processes of the brain (17). As is evident from the data obtained in this study, a 100% survival rate was directly related to the duration of hypothermia. With this treatment 5 min after onset of focal ischemia, mortality apparently did not depend on the infarction size. The application of hypothermia obviously retarded the sequence of pathological events most probably due to a temperature-dependent reduction or delay of biochemical reactions. However, this transient "protective" effect of hypothermia was not maintained after rewarming to normothermia. The outcome after different periods of hypothermia was within the same range as in untreated animals indicating that under reestablished normothermic conditions, the deteriorating effects of regional ischemia could not be prevented but delayed.

The metabolic inhibitory effect of barbiturates has been related to the reduced rate of energy reserve utilization in the cerebral cortex after complete ischemia by decapitation (7). Under hypoxemic conditions, the metabolic inhibition was interpreted as the consequence of mild hypothermia during barbiturate anesthesia (19). If the protective effect of barbiturates is due to reduced cerebral metabolism, one should expect improved recovery of the brain after global ischemia by preventing the impairment of postischemic hypermetabolism. The results obtained, however, are not very encouraging (25). Reports on experimental focal ischemia indicate beneficial effects of barbiturates applied before or after the onset of the ischemic impact (8,9,14,20,22,28). It has been suggested before that other mechanisms of barbiturate action than metabolic inhibition may play a role for the better outcome after focal ischemia. A tissue concentration of 10 to 20 μg pentobarbital per gram brain tissue, which is reached after an intraperitoneal dose of approximately 10 mg/kg pentobarbital, causes 50 - 60% of total depression of glucose utilization. At higher tissue levels of pentobarbital, the remaining proportion of glucose consumption apparently is not sensitive to barbiturate depression (5). Halothane, for instance, reduces $CMRO_2$ as well (26), but no protective effect was found for focal ischemia. On the contrary, infarct size increases under deep halothane anesthesia, which suggests that the metabolic inhibition per se is not the only protective effect under these experimental conditions (22). There are indications that the mode of barbiturate action also involves a decrease of edema formation (20), reduction of intracranial pressure (24), and free radical scavenging (6) in regional ischemia.

As is evident from our data, immediate barbiturate treatment at a dose of 50 mg/kg increases the survival rate to 57% as compared to 28% in untreated controls. About 35% of the animals exhibit severe symptoms and it is likely that the extent of ischemic brain damage is larger than in the rest of the gerbils which reveal mild or no neurological deficits. It has been reported before that the reduction of blood flow by barbiturates alters intracranial pressure through a decrease of intracranial blood volume (21). These hemo-dynamic changes were interpreted to be the primary factors for the protective mechanisms of barbiturates in focal ischemia. When additional intracranial reserve space is induced by barbiturates which may be supported by reduced edema formation, inherniation due to intracranial hypertension might be prevented in animals with small sized ischemic brain lesions. However, in animals with large brain infarcts, intracranial reserve spaces are exhausted rapidly and barbiturates appear to have no influence on the prevention of intracranial hypertension.

The absence of beneficial effect of barbiturates on the survival rate for all doses of pentobarbital applied 5 min after onset of vascular occlusion, however, implies that mechanisms of barbiturate action other than only hemodynamic changes are present. During this "unprotected" period of time, tissue damage may occur as the consequence of biochemical reactions in the ischemic tissue. It has been suggested that free radical scavenging may be disturbed thus leading to lipid-free radical reactions in mitochondrial membranes (6). After the initiation of cerebral tissue damage, no evidence is obtained in the studies reported here that a delayed dose of barbiturates positively influences the existing pathological changes in terms of improved survival rates.

It is concluded that the influence of barbiturates on the improved outcome after onset of focal ischemis is time-dependent. In this animal model, the positive effect on the survival rate can only be demonstrated when barbiturates are given immediately after vascular occlusion at a high dose. The reduction of intracranial pressure due to the decrease of blood flow may contribute to the better outcome after onset of critical ischemia as well as preventing ongoing pathophysiological processes in ischemic tissue. On the other hand, during hypothermia it was possible to delay progressive tissue damage thus improving the survival rate after onset of regional cerebral ischemia. Hypothermia, therefore, may be of advantage to prolong the duration of ischemia after which effective therapeutic treatment may be started.

References

1. Arnfred I, Secher O (1962) Anoxia and barbiturates: Tolerance to anoxia in mice influenced by barbiturates. Arch Int Pharmacodyn Ther 139:67-74

2. Blöink M, Hossmann V, Hossmann K-A (1979) Treatment of experimental infarcts following middle cerebral artery occlusion in cats. In: Bès A, Géraud G. Eds. Circulation cérébrale. Excerpta Medica, Toulouse, pp 85-87

3. Bosma H-J, Paschen W, Hossmann K-A (1981) Cerebral ischemia in gerbils using a modified vascular occlusion. In: Meyer JS, Lechner H, Reivich M, Ott EO, Aranibar A. Eds. Cerebral Vascular Disease 3. Excerpta Medica, Amsterdam, pp 280-285

4. Carlsson C, Hägerdal M, Siesjö BK (1976) Protective effect of hypothermia in cerebral oxygen deficiency caused by arterial hypoxia. Anesthesiology 44:27-35

5. Crane PD, Braun LD, Cornford EM, Cremer JE, Glass JM, Oldendorf WH (1978) Dose-dependent reduction of glucose utilization by pentobarbital in rat brain. Stroke 9:12-18

6. Flamm ES, Demopoulos HB, Seligman ML, Ransohoff J (1977) Possible molecular mechanism of barbiturate-mediated protection in regional cerebral ischemia. Acta Neurol Scand (Suppl 64) 56:150-151

7. Gatfield PD, Lowry OH, Schulz DW, Passonneau JV (1966) Regional energy reserves in mouse brain and changes with ischemia and anaesthesia. J Neurochem 13:185-195

8. Hankinson HL, Smith AL, Nielsen SL, McDonald LW, Youmans JR (1974) Effect of thiopental on focal cerebral ischemia in dogs. Surg Forum 25:445-447

9. Hoff JT, Smith AL, Hankinson HL, Nielsen S (1975) Barbiturate protection from cerebral infarction in primates. Stroke 6:28-33

10. Kogure K, Alonso OF (1978) A pictorial presentation of endogenous brain ATP by a bioluminescence method. Brain Res 154:273-284

11. Levine S, Payan H (1966) Effects of ischemia and other procedures on the brain and retina of the gerbil (Meriones unguiculatus). Exp Neurol 16:255-262

12. McGraw CP (1977) Experimental cerebral infarction. Effects of pentobarbital in Mongolian gerbils. Arch Neurol 34:334-336

13. Michenfelder J, Theye R (1970) The effects of anesthesia and hypothermia on canine cerebral ATP and lactate during anoxia produced by decapitation. Anesthesiology 33:430-439

14. Michenfelder JD, Milde JH, Sundt TM Jr (1976) Cerebral protection by barbiturate anesthesia: use after middle cerebral artery occlusion in Java monkeys. Arch Neurol 33:345-350

15. Mies G, Niebuhr I, Hossmann K-A (1981) Simultaneous measurement of blood flow and glucose metabolism by autoradiographic techniques. Stroke 12:581-588

16. Paschen W, Niebuhr I, Hossmann K-A (1981) A bioluminescence method for the demonstration of regional glucose distribution in brain slices. J Neurochem 36:513-517

17. Rosomoff L (1959) Protective effects of hypothermia against pathological processes of the nervous system. Ann NY Acad Sci 80:475-486

18. Sakurada O, Kennedy C, Jehle J, Brown JD, Carbin GL, Sokoloff L (1978) Measurement of local cerebral blood flow with iodo(^{14}C)antipyrine. Am J Physiol 234:H59-H66

19. Siesjö BK, Nillson L (1971) The influence of arterial hypoxemia upon labile phosphates and upon extracellular lactate and pyruvate concentrations in the rat brain. Scand J Clin Lab Invest 27:83-96

20. Simeone FA, Frazer G, Lawner P (1979) Comparative effects of barbiturates and hypothermia. Stroke 10:8-12

21. Smith AL, Wollman H (1972) Cerebral blood flow and metabolism: Effects of anesthetic drugs and techniques. Anesthesiology 36:378-400

22. Smith AL, Hoff JT, Nielsen SL, Larson CP (1974) Barbiturate protection in acute focal cerebral ischemia. Stroke 5:1-7

23. Sokoloff L, Reivich M, Kennedy C, Des Rosiers MH, Patlak CS, Pettigrew KD, Sakurada O, Shinohara M (1977) The (2-^{14}C) deoxyglucose method for the measurement of local cerebral glucose utilization: theory, procedure and normal values in the conscious and anesthetized albino rat. J Neurochem 28:897-917

24. Sondergard W (1961) Intracranial pressure during general anesthesia. Dan Med Bull 8:18-25

25. Steen PA, Milde JH, Michenfelder JD (1978) Cerebral metabolic and vascular effect of barbiturate therapy following complete global ischemia. J Neurochem 31:1317-1328

26. Theye RA, Michenfelder JD (1968) The effect of halothane on canine cerebral metabolism. Anesthesiology 29:1113-1118

27. Welsh FA, Rieder W (1978) Evaluation of in situ freezing of cat brain by NADH fluorescence. J Neurochem 31:299-309

28. Yatsu FM, Diamond I, Graziano C, Lindquist P (1972) Experimental brain ischemia: Protection from irreversible damage with a rapid-acting barbiturate (methohexital). Stroke 3:726-732

A Dog Model to Evaluate Post-Cardiac Arrest Neurological Outcome

A. Mullie[1], K. Vandevelde[1], H. van Belle[2], A. Jagenau, J. van Loon[2], C. Hermans[2], and A. Wauquir[2]

1 St. Jan Hospital, Critical Care Medicine Department, Brugges, Belgium
2 Janssen Pharmaceutica, Departments of Biochemistry and Neuropharmacology, 2340 Beerse, Belgium

Introduction

The success rate of cardiopulmonary resuscitation (CPR) in circulatory arrest (CA) even when performed by the most experienced emergency medicine groups is still less than 15%. The limit to final outcome is not the restoration of spontaneous circulation (ROSC) itself, but the events occurring following reperfusion of the brain by acid and toxic blood during CPR. The aim was to develop a dog model of ischemia in order to study the potential of physiological and pharmacological methods for restoring normal brain function after prolonged circulatory arrest.

Materials and methods

Beagle dogs about one year of age and weighing 10 - 15 kg were the subjects of the study. The cerebral insult consisted of 12 1/2 min of total circulatory arrest by electrical cardiac fibrillation in awake animals, followed by a standardized CPR period of less than 5 min. This insult resembles sudden clinical death from ventricular fibrillation (VF).

During the 12 1/2 min period of CA, the animal is intubated, an intra-arterial catheter is positioned in the aortic arch under fluoroscopic control, and a central venous line is inserted via the femoral vein. All animals are resuscitated within 5 min by external cardiac compression (ECC), intermittent positive pressure ventilation (IPPV), intra-aortic adrenaline (0.1 mg/kg) and sodium bicarbonate (2 meqV/kg) administration and electrical defibrillation. After resuscitation, the animals are subsequently kept under intensive monitoring and cardiopulmonary stabilization by all currently known intensive care techniques (i.e. IPPV, frequent blood gas analysis, fluid and dopamine titration, PEEP) for 6 h (Fig. 1).

Neurological recovery is evaluated by three parameters: a) a clinical coma score (Table 1) was determined repeatedly during the last 3 h of the experiment, b) inorganic phosphate and c) creatinine phosphokinase-isoenzyme BB fraction in cerebrospinal fluid at 1,3 and 6 h post CA was determined.

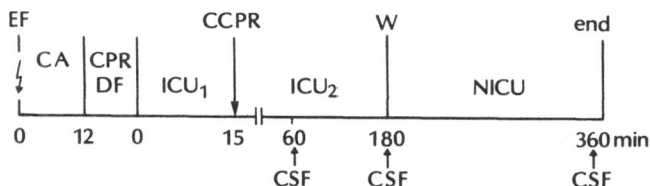

EF = electrical fibrillation
CA = circulatory arrest
CPR = cardiopulmonary resuscitation
DF = defibrillation
CCPR = „brainresuscitating" drug administration
ICU_1 = IPPV-VET CO_2-4
      titration of rheomacrodex (CVP~10mm Hg)dopamine,
      pancuroniumbromide, sodiumbicarbonate.
ICU_2 = IPPV, dopamine, EEG
W = weaning from IPPV
NICU = intensive neurological observation score

Fig. 1
Dog model to evaluate
post-cardiac arrest brain
damage

Table 1 Clinical neurological score

1. eye movement	0-9
2. skeletal muscle response	0-8
3. cranial nerve reflexes	0-9
4. pupillary reflex	0-8
5. respiratory pattern	0-9
6. ability to wean from IPPV	0.7

cerebral death = 0 - normal = 50 points

Results and discussion

Definite progress in the problem of the high failure rate of CPR after CA can only be
researched in an animal model, closely resembling sudden clinical death (1,2,3,5). Our model
imitates the clinical problem of prolonged CA by ventricular fibrillation in an awake animal
(1,2): a prolonged CA to over 10 min, a short standardized CPR and finally a period of
intensive care by cardiopulmonary stabilization techniques. Variability in CPR performance,
often related to some degree of electromechanical dissociation (2) after prolonged CA, is
prevented by intra-aortic adrenaline administration. Resuscitation is no problem in our

model, the immediate rate of success of our CPR is almost 100%; occasionally, 1 - 2 extra doses of adrenaline are required.

The problems in the model arise with the development of "post-resuscitation disease". The complex pathophysiological state following resuscitation requires a control of the different cardiopulmonary and metabolic variables (i.e. bleeding, hemolysis, sepsis, hyperthermia). The most impressive finding during our experience with this model is the tremendous importance of cardiopulmonary stability in the early post CA period as the determinant for final neurological outcome. Mainly in more prolonged CPR animals (4-5 min), we observe a shock-lung-like syndrome (hypoxia, acidosis, increased dopamine requirements) 2 - 4 h post ROSC. This phenomenon almost certainly predicts a lower coma score (Table 2). Blood pressure stabilization itself, which sometimes requires very high amounts of dopamine, does not prevent the deleterious neurological consequences of the secondary cardiopulmonary lesion.

Brain damage post CA has been attributed to several pathophysiological events, i.e. multifocal postischemic hypoperfusion, abnormally high metabolism persistent tissue acidosis, potentially injurious chemical compounds in and around brain cells (fatty acids, prostaglandins, neurotransmitters, free radicals).

Table 2   Coma score in dogs after 12 1/2 min CA, less than 5 min CPR, followed by 6 h intensive care with and without "shock-lung"-like syndrome

shock-lung-like present	shock-lung-like absent
12	27
12	29
5	45
8	46
5	44
7	48
4	40
2	48
12	41
0	42
8	46
12	45
7	45
mean 7.2 (0-12)	mean 42 (27-48)

Table 3 CSF of dogs post 2 1/2 min CA in which less than 5 min CPR was needed

CSF-Pi	neurological outcome score	CSF-CPK$_{BB}$
0.81	4	-
0.98	8	30
1.14	12	6
1.18	14	20
0.95	27	-
0.98	29	2
0.66	44	2
0.72	45	2
0.61	46	4
0.70	48	-

In the present study, three parameters which should reflect developing brain pathology were measured: 1) measurement of CSF inorganic phosphate (reflecting energy metabolism); 2) creatine phosphokinase isoenzyme BB (reflecting brain cell breakdown); 3) a clinical neurological score (Table 1). The results shown in Table 3 compare the three parameters.

CSF analysis was performed in 1ml of fluid drawn from the cisterna magna. There appeared to be no direct relationship between CSF-Pi, neurological score and CSF-CPK$_{BB}$. These parameters do not therefore appear to be an early biochemical index of later neurological deficits.

Under our experimental conditions, dogs were subjected to a 6 h intensive care period post CA. Whether the neurological state at this time accurately reflects final neurological outcome might be questioned. However, from pilot studies in our model with 24 h life support and because of the reproducibility of neurological outccme in the control animals, we presume that an evaluation after 6 h is predictive. In preliminary experiments, the therapeutic effects of a low i.v. dose of flunarizine (4) (0.1 mg/kg), started 15 min after resuscitation and titrated over 50 min, were evaluated. Some improvement was obtained in dopamine-independent dogs, though the neurological outcome in control dogs appeared also relatively beneficial. Further pilot studies will, however, be necessary in order to determine the optimal dose and time of flunarizine administration with respect to improved neurological outcome.

Conclusions

1. This dog model is a close correlate to sudden ventricular fibrillation in humans.
2. Intra-aortic drug administration during CPR is an effective method to prevent electromechanical dissociation.
3. Prevention of the secondary cardiopulmonary events is an important goal in the immediate post-CPR period.
4. The determination of Pi and CPR_{BB} in the CSF has failed as an early index of possible neuronal outcome.

References

1. Hermans C, De Reese R, Van Loon J, Loots W, Jagenau THM (1982) A cardiac arrest model in rats for evaluating the antihypoxic action of flunarizine. Eur J Pharmacol 81:137-140

2. Mullie A, Hermans C, Vandevelde K, Wauquier A (1981) Resuscitability with brain protective drugs during cardiopulmonary resuscitation in dogs. Crit Care Med 93:183

3. Safar P, Bleyaert A, Nemoto EM, Mossy J, Snyder JV (1978) Resuscitation after global brain ischemia-anoxia. Crit Care Med 6:215-227

4. Wauquier A (1982) Brain protective properties of etomidate and flunarizine. J Cereb Blood Flow Metab (Suppl 1) 2:S53-S56

5. White BC, Gadzinski DS, Hoehner PJ, Krome C, Hoehner T, White JD, Trombley JH (1982) Correction of canine cerebral cortical blood flow and a vascular resistance after cardiac arrest using flunarizine, a calcium antagonist. Ann Emer Med 11:119-126

Cortical Glucose and Energy Metabolism During Complete Cerebral Ischemia and After Recovery

C. Krier[1] and S. Hoyer[2]

1 Klinikum der Universität Heidelberg, Zentrum Chirurgie, Abteilung für Anästhesiologie, Im Neuenheimer Feld, 6900 Heidelberg 1, FRG
2 Institut für Pathochemie und Allgemeine Neurochemie, Zentrum Pathologie, Universität Heidelberg, Im Neuenheimer Feld 220-221, 6900 Heidelberg 1, FRG

After cardiac arrest producing brain injury, cerebral resuscitation measures have to be initiated during cardio-pulmonary resuscitation and continued in the recovery period in order to limit brain damage after complete cerebral ischemia.

Brain-oriented life support can improve the neurological outcome of patients undergoing brain injury produced by circulatory arrest (3,4,14,15).

Global brain ischemia with reperfusion is followed by multifocal necrosis and a multitude of secondary changes such as vasoparalysis, tissue edema, membrane damage, neurotransmitter failure and transient hypermetabolism leading to cerebral dysfunction and neuronal death (6,7,12).

Brain resuscitation and protection measures with thiopentone and similar drugs reported to depress brain metabolism is well documented in focal brain ischemia, hypoxia and hypoperfusion states, although the exact mechanism of action remains unknown (7,10,18). In contrast, brain resuscitation after complete global ischemia such as occurs after cardiac arrest with brain metabolism depressing drugs is still controversial.

In order to define the utility of large doses of drugs depressing brain metabolism, we investigated the effect of thiopentone, etomidate and gamma-hydroxy-butyrate on carbohydrate and energy metabolism of rat cortex in an experimental model of complete reversible ischemia followed by a recovery period.

Methods

Complete global brain ischemia was produced in one year old male Wistar rats by temporary looping of both carotid and vertebral arteries at the aortic arch after thoracotomy. Mean arterial blood pressure was lowered during the ischemia period to 30 mmHg in order to prevent collateral supply of the brain and to assure completeness of ischemia.

Anesthesia was induced with 3.0% halothane and nitrous oxide/oxygen 70:30; at the end of surgical preparation, halothane was discontinued and the steady state only started after halothane had been monitored to be at 0.03 Vol% or lower.

The rats were paralyzed with pancuronium bromide and mechanically ventilated with 70% nitrous oxide and 30% oxygen supply. Femoral artery and vein were cannulated for blood samplings and pressure recording. Temperature was monitored subcutaneously.

After a steady state of 20 min of normothermia, arterial normotension, normoxia and normocapnia, complete cerebral ischemia was performed for 15 min. Recirculation/re-oxygenation period followed for 60 min in the spontaneous recovery group and for 60 min in the treated group under the steady state conditions mentioned above (Fig. 1). The brains were frozen in situ by liquid nitrogen in a funnel formed from the scalp. Brain cortex was analyzed for the concentration of glucose amd lactate and energy-rich compounds using sensitive standard enzymatic methods (8).

After reopening of the looped arteries, thiopentone, etomidate and gamma-hydroxy-butyric acid were administered by bolus followed by continuous infusion for 60 min in the dose of 20 mg . kg^{-1} bolus and 20 mg . kg^{-1} . h^{-1} for thiopentone, 7.5 mg . kg^{-1} bolus and 30 mg . kg^{-1} . h^{-1} for etomidate and 363 mg . kg^{-1} bolus and 18 mg . kg^{-1} . h^{-1} for gamma-hydroxy-butyric acid. The dosage of the mentioned drugs had been evaluated in previous experiments in order to produce and maintain a burst-suppression EEG during the entire recovery period, thus corresponding to the large loading doses recommended in clinical trials (5). The rats were divided into nine groups of six rats each as shown in Fig. 2.

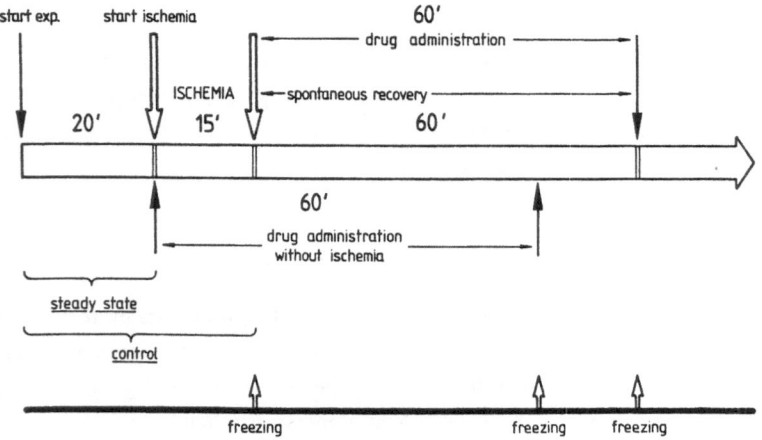

Fig. 1

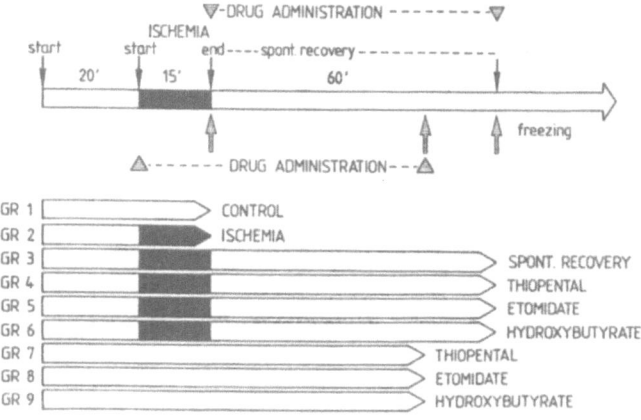

Fig. 2

Results

Figure 3 and Table 1 show the effect of complete ischemia and of the respective recovery periods on the concentrations of glucose and lactate and high-energy phosphates in percentage changes of control values.

The cerebral ischemia of 15 min duration produced a sharp drop in glucose concentration to $0.078 \mu mol.g^{-1}$ and a dramatic 12-fold increase in lactate production to around $14 \mu mol.g^{-1}$. An almost complete depletion of tissue content of creatine phosphate, ATP and ADP, but a marked augmentation of AMP was measured (Fig. 3, Table 1).

In the spontaneous reovery group, we found an around 3-fold increase of glucose as compared to control conditions. Lactate was also increased to more than 6-fold as compared to controls, but was around 50% of that concentration found after complete cerebral ischemia without reperfusion period. The energy-rich phosphates had nearly normalized.
After thiopentone, etomidate and gamma-hydroxy-butyrate administration under control conditions, brain glucose was found to be slightly elevated and lactate concentration dropped below control values. There were no marked changes in the concentrations of energy-rich compounds (Table 2).

Thiopentone, etomidate and gamma-hydroxy-butyrate did not markedly influence the increased glucose concentration as was found under spontaneous recovery and as compared to the controls. The increase in lactate production remained enhanced under these drugs, but it was less augmented under gamma-hydroxy-butyrate. The drug effect on the high-energy phosphates was similar: a tendency to normalization (Fig. 3, Table 1).

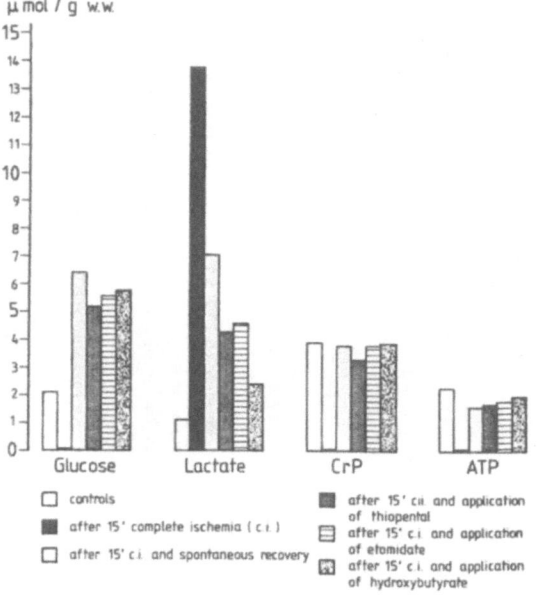

μ mol / g w.w.

Glucose   Lactate   CrP   ATP

□ controls

■ after 15' complete ischemia (c.i.)

□ after 15' c.i. and spontaneous recovery

■ after 15' c.i. and application of thiopental

▦ after 15' c.i. and application of etomidate

▨ after 15' c.i. and application of hydroxybutyrate

Fig. 3

Discussion

The predominant factor of tissue damage during ischemia is production of lactate and tissue acidosis, respectively, on the one hand and inhibition of respiration on the other hand. The rise of lactate, partly due to ischemia and the consequent fall in tissue pH has been implicated as a major mechanism of brain damage leading to brain edema, thus affecting the postischemic recovery (17). More recently, MacMillan demonstrated that ischemia, in which the tissue lactate levels are below $20 \, \mu mol \, g^{-1}$, result in a paucity of early histological changes since Na^+, K^+-ATPase activity remains unaffected (9).

It has been suggested hence that a high degree of lactate accumulation in the brain tissue seriously impairs recovery in the postischemic period, and that metabolites of glycolysis are among other factors responsible for the fall in tissue pH (11,13,17).

Accordingly, reduction of glucose uptake and inhibition of lactate production is suggested to be a protective mechanism - among others - and a drug able to diminish hypermetabolism of glucose and lactate production should have a resuscitating effect after brain ischemia.

The data of our study could not clearly confirm the evidence of brain protection after complete global ischemia by means of metabolism-depressing drugs - at least for the acute phase of recovery and for the dosage used in our model. Glucose augmentation was similar in

Table 1   Cortex content of ATP, ADP and AMP under ischemia, spont, recovery
and under drug treatment following 15'-ischemia

	ATP	ADP	AMP
Control	2.26	0.275	0.022
Ischemia 15'	0.06	0.274	1.52
Isch.+ spont. recovery	1.63	0.210	0.025
Ischemia + thiopentone	1.73	0.297	0.051
Ischemia + etomidate	1.81	0.247	0.034
Ischemia + gamma-hydroxy-butyrate	2.04	0.250	0.027

Table 2   Cortex content of glucose, lactate and energy-rich-phosphates under
thiopentone, etomidate and gamma-hydroxy-butyrate administration under
control conditions

μmol g^{-1} w.w.	Glucose	Lactate	ATP	ADP	AMP
Control	1.940	1.227	2.271	0.286	0.056
Thiopentone	2.978	0.925	2.285	0.321	0.040
Etomidate	2.484	0.819	2.300	0.302	0.061
Gamma-hydroxy-butyrate	2.810	0.7830	2.490	0.275	0.027

the drug-treated groups as compared to the spontaneous recovery group (Fig. 3). Whether
the obviously decreased lactate production under the drugs administered as compared to the
spontaneous recovery group would have a resuscitating effect after global brain ischemia
has to remain open as yet, since the concentration of lactate remains abnormally elevated.

For thiopentone and etomidate in the mentioned dosage, a beneficial effect in the acute
phase after complete brain ischemia could also not be demonstrated by means of changes in
energy-rich compounds.

Other mechanisms of thiopentone and etomidate may have a beneficial effect in the complicated pathophysiology of brain failure. On the level of carbohydrate and energy metabolism, a convincing beneficial effect could not be postulated for both drugs. It has nevertheless to be emphasized that the control group is not to be considered as a strictly non-treated group but as a group treated by immobilization, normotension, mechanical ventilation and nitrous oxide anesthesia.

The significant decrease of lactate concentration in the gamma-hydroxy-butyric acid-treated group may possibly indicate a beneficial effect of the drug at the level of lactate formation, since the tissue concentration was reduced from 7-fold increase after spontaneous recovery to 2.4-fold increase in the drug-treated group. In summary, convincing evidence of the beneficial effect of metabolism-depressing drugs could not be postulated for the acute postischemic period in our model, except perhaps for gamma-hydroxy-buteric acid.

The suggestion that large doses of barbiturates should be administered after complete brain ischemia thus remains controversial; it has to be emphasized that brain resuscitation measures have to consider the extremely multi-factorial pathophysiology of brain failure.

Focussing the resuscitation measures on the beneficial effect of a single drug at whatever dosage level chosen would neglect this fact and seriously compromise the chance of recovery with other brain-oriented life-support techniques such as hyperventilation, normo-perfusion and restitution of the normality of the 'milieu intérieur'.

References

1. Ames A, Wright RL, Kowada M, Thurston JM, Majno G (1968) Cerebral ischemia II. The no-reflow phenomenon. Am J Path 52:437-458

2. Black KL, Weidler DJ, Jallad NS, Sodeman TM, Abrams GD (1978) Delayed pento-barbital therapy of acute focal cerebral ischemia. Stroke 9:245-249

3. Bleyaert A, Nemoto EM, Safar P, Stezoski W, Mickell J, Moossy J, Rao GR (1978) Thiopental amelioration of brain damage after global ischemia in monkeys. Anesthesiology 49, 390-398

4. Bleyaert A, Safar P, Nemoto E, Moossy J, Sassano J (1980) Effect of postcirculatory-arrest life-support on neurological recovery in monkeys. Crit Care Med 8:153-156

5. Breivik H, Safar P, Sands P, Fabritius R, Lind B, Lust P, Mullie A, Orr M, Renck H, Snyder JU (1978) Clinical feasibility trials of barbiturate therapy after cardiac arrest. Crit Care Med 4:228-244

6. Jost U, Bortel HJ, Schmitt H, Hoyer S (1980) Beeinflussung der cerebralen Glycolyse und Atmungskettenoxidation (dem neuronalen energieliefernden Stoffwechsel) durch Thiopental, Flunitrazepam und Etomidate im steady state einer standardisierten Inhalationsnarkose. Anaesthesist 29:12-17

7. Levy DE, Brierley JB (1979) Delayed pentobarbital administration limits ischemic brain damage in gerbils. Ann Neurol 5:59-64

8. Lowry OH, Passonneau JV (1972) A flexible system of enzymatic analysis. Academic Press, New York

9. MacMillan V (1982) Cerebral Na^+, K^+-ATPase activity during exposure to and recovery from acute ischemia. J Cereb Blood Flow Metabol 2:457-465

10. Michenfelder JD, Milde HJ, Sundt TM (1976) Cerebral protection by barbiturate anesthesia use after middle cerebral artery occlusion in Java monkeys. Arch Neurol 33:345-350

11. Myers RE (1979) A unitary theory of causation of anoxic and hypoxic brain pathology. In: Fahn S, Davis JN, Powland LP. Eds. Advances Neurol. Vol. 26. Raven Press, New York, pp 195-213

12. Nemoto EM (1978) Pathogenesis of cerebral ischemia-anoxia. Crit Care Med 6:203-214

13. Rehncrona S, Rosén I, Siesjö BK (1981) Brain lactic acidosis and ischemic cell damage. 1. Biochemistry and neurophysiology. J Cereb Blood Flow Metabol 1:297-311

14. Safar P (1981) Cardiopulmonary resuscitation. W.B. Saunders Co., Philadelphia Toronto London

15. Safar P, Bleyaert AL, Nemoto EM et al (1978) Resuscitation after global brain ischemia. Crit Care Med 6:215-217

16. Siesjö BK (1978) Brain Energy Metabolism. Wiley, Chichester New York Brisbane Toronto

17. Siesjö BK (1981) Cell damage in the brain: A speculative synthesis. J Cereb Blood Flow Metabolism 1:155-185

18. Steen PA, Milde JH, Michenfelder JD (1978) Cerebral metabolic and vascular effects of barbiturate therapy following complete global ischemia. J Neurochem 31:1317-1324

Monitoring of Cerebral Ischaemia in Man

L. Symon

Gough Cooper Department of Neurological Surgery, Institute of Neurology,
Queen Square, London WC1, United Kingdom

Event related potentials in the central nervous system in response to peripheral nerve stimulation represent a valuable index of the mechanisms determining human somato-sensory processes. They are of value in the diagnosis of lesions affecting the afferent pathways (5,10,11,16,19) or in the prediction of outcome (7,15). In previous work, the relationships between the amplitude of SEP components considered to originate in cerebral cortex and the local tissue blood flow have been established in the baboon (2). We have found and decsribed a threshold relationship, normal SEP amplitude being sustained only when flow is greater than about 18 ml/100g/min. We have extended this experimental investigation into a study of sensory evoked potential (SEP) in man in relation to the ischaemic events which may be associated with aneurysmal subarachnoid haemorrhage, and also to continuous monitoring of the electrical activity of the cortex during surgical operations.

The methods used were those of Hume and Cant (6), voltage pulses of 0.15 ms being delivered at a rate of three per second to silver cup electrodes over the median nerve with the cathode approximately 3 cm proximal to the anode. The presence of a thumb twitch indicated an adequate level of stimulation (7).

Recording electrodes were placed over the right and left central scalp region (C4/C3 international 10-20 system) and over the second cervical vertebra. The reference electrode was placed on the mid-forehead (FPZ). An average of 256 or 512 responses were recorded from C2 and the contra-lateral sensory cortex using a variety of recording systems. The main measurement which we made was that of central conduction time determined by subtracting the latency of N14, the arrival of the impulse at the dorsal column nuclei recorded at C2 from the arrival of the post-synaptic volley at the contralateral cortex (N20). The CCT represents the central pathways only and is independent of peripheral nerve or spinal cord tracts.

We have studied 90 patients during the course of the evolution of aneurysmal subarachnoid haemorrhage following admission to hospital, and 33 patients during operation in which 38 aneurysms were clipped in a total of 34 operations.

Results

Central conduction time measurements were found to be to some extent predictive of the likely outcome of the subarachnoid haemorrhage. Thus in the 90 patients examined pre-operatively and at serial intervals during the evolution of the subarachnoid haemorrhage and following operation, we found that there was a significant difference between subarachnoid haemorrhage graded as I to III in the Hunt and Hess system (8), and those of grade IV. Of rather more importance than this, however, was the presence of central conduction time more than two standard deviations above that of control volunteers in any hemisphere. Thirteen subarachnoid haemorrhage patients without aneurysm and 12 normal subjects had a central conduction time of 5.4 + 0.4 ms. Taking central conduction time of more than 2.5 standard deviations from the norm as a guide, we have found a significant difference between those patients who had a good outcome and those who had a poor outcome, that is

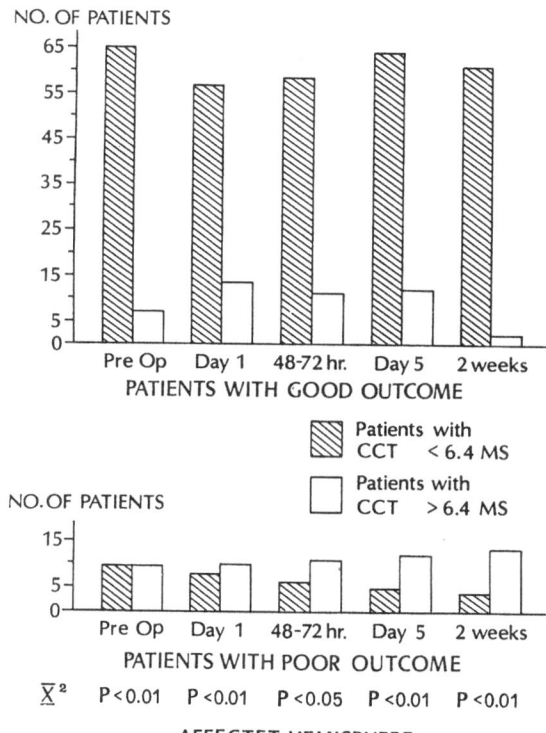

Fig. 1   Histogram distribution of patients at various stages following subarachnoid haemorrhage with and without significant prolongation of central conduction time (normal, mean plus two standard deviations). There is a significant difference in the distribution as shown, more patients with prolonged conduction time having a poor outcome

who were at two months still in care or had died (Fig. 1). These outcome data were to some extent independent of the pre-operative grade, in that more patients with grade III and IV with normal conduction time fell into the good outcome group and patients in those grades with a conduction time of more than 6.4 ms fell into the poor outcome grades. It would appear, therefore, that this technique is of some use in differentiating patients in grade III or IV from one another. It may therefore add to one's clinical interpretation of the advisability of surgery on patients with subarachnoid haemorrhage.

Operative recording

Recordings throughout operation were carried out in 33 patients. Pre-operative recording in these patients gave a CCT before operation of 5.8 ± 0.5 ms with a range of 5.2 - 7 ms. The pre-operative recordings varied from 5.7 ± 0.4 ms in grade I patients to 6.0 ± 0.5 ms in grade IV patients. There was a significant difference only between grade IV patients and the control group (P<0.01). Induction of anaesthesia with thiopentone followed by intubation slightly slowed the conduction time. The standard form of maintenance of anaesthesia thereafter, inhalational halothane in concentrations varying from 0.5 to 2% resulted in moderate hypotension and in all 33 cases CCT was prolonged (Fig. 2). The average value after the attainment of a stable level of anaesthesia was 6.7 ± 0.7 ms (P < 0.01 compared with pre-operative). Blood pressure was significantly reduced from a mean SBP of $87.9 \pm$ mmHg to 70.1 ± 12.1 mmHg after a stable level of halothane concentration had been obtained (P < 0.01). CCT was found occasionally useful in controlling retraction. In four patients, conduction time in the retracted hemisphere exceeded that in the control hemisphere by more than 0.6 ms which is the maximum interhemispheric difference which we found in our control group plus 2.5 standard deviations. During the dissection of the aneurysm and its parent vessels, CCT again gradually prolonged. The application of an aneurysm clip to 13 aneurysms in 12 of the 33 patients produced definite prolongation of CCT varying from 0.4 to 3 ms. Temporary clips were applied to major vessels in 15 patients, and in nine of these there was no detectable change in CCT despite clips to terminal carotid, middle cerebral or anterior cerebral arteries. In the other six, however, there was an immediate and significant prolongation of CCT within 3 min by up to 3 ms, and in one case the N20 could not be identified after the application of temporary clips to the posterior communicating artery and terminal carotid artery above and below the neck of a complex posterior communicating artery aneurysm. The removal of temporary clips resulted in the recovery of conduction velocity to the immediately preceding level within 15 minutes except in the single case of triple vascular occlusion where a recognizable potential appeared within 8 min but the potential did not approach the preceding level until 40 min had elapsed.

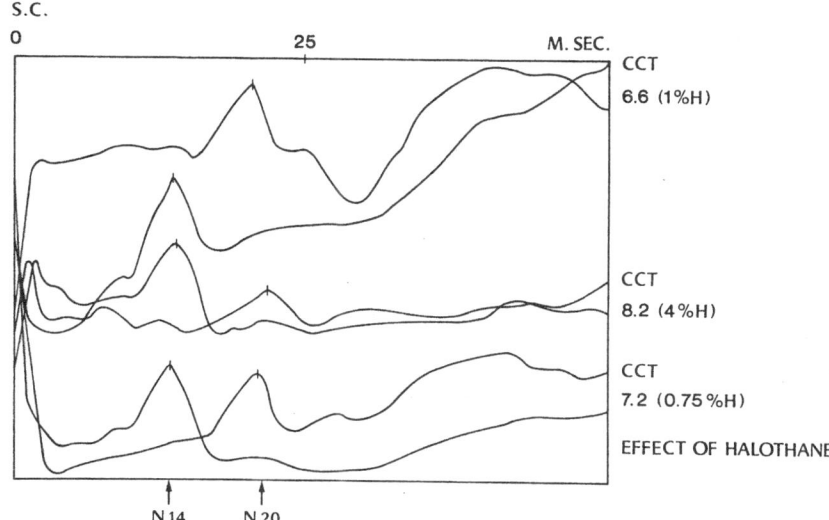

Fig. 2   SEP tracings during operation, recording from C2 (N14) and the somatosensory cortex (N20). The effect of varying concentrations of halothane on the conduction time are shown

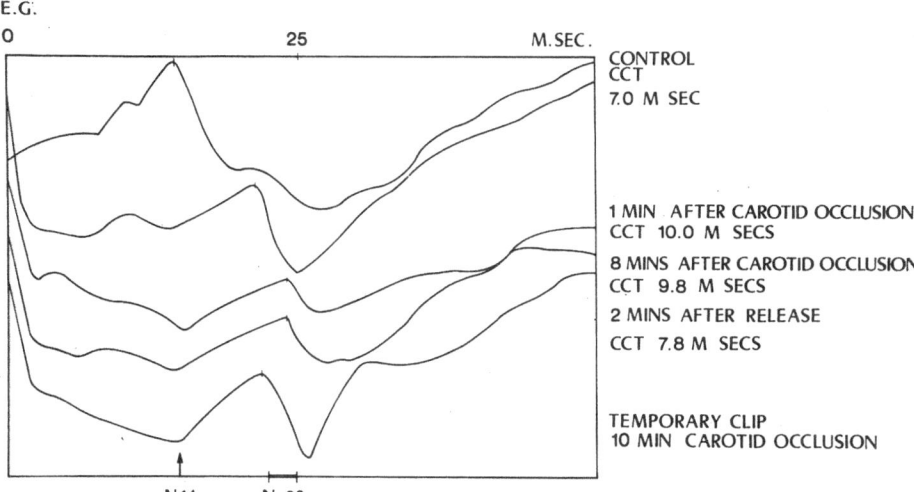

Fig. 3   The effects of temporary occlusion of the internal carotid artery on central conduction time. A single control cervical record is shown in the upper trace, and serial somatosensory cortex recordings (from scalp) shown below

Amplitude measurements

We have not found measurement of the amplitude of SSEP in the ward or theatre of such considerable value. In general, increases in conduction time were associated with a decrease in amplitude although this was not invariably the case as in some instances, as conduction time prolonged, there was an apparent increase in amplitude early during the development of ischaemia. This has also been seen in experimental animals.

Alterations in conduction velocity could be detected at various stages before detactable decline in amplitude, and comparison of the N20 and N14 peak heights revealed no extra information except during occlusion. The same was true of comparison between the hemispheral evoked response heights. Only in the circumstance of individual hemisphere retraction or occlusion of the vessel on one side did the amplitude of the two traces differ significantly. Amplitude measurements were further complicated by incidental noise to a much greater extent than the measurement of conduction velocity.

Discussion

Monitoring patients during developing cerebral ischaemia may be episodic or continuous. In episodic monitoring we may cite for example, inhalational cerebral blood flow techniques either by inhaled radio-active Xenon or by the use of non-radioactive Xenon with CAT scanning. These are unfortunately time consuming and difficult to repeat frequently in seriously ill patients.

For continuous monitoring, we must turn either to implanted electrode analysis of CBF as by the technique of hydrogen clearance which may be performed frequently but has the considerable disadvantage of requiring electrode implantation, or we may use continuous electrical recording. While the EEG is useful in this regard particularly with the use of modern averaging techniques, the evoked electrical activity is of particular interest since it has been closely tied to the adequacy of perfusion of tissue. We may therefore use evoked electrical activity in the diagnosis of the site and severity of lesions, either in the brain stem or cortex, and this can also provide information about the functional state of the nervous system which is hard to obtain by other means in patients who may be comatose or under treatment (1,3,4,13,14,17,18) with a muscle relaxant or extensive sedation. Evoked response recording is of value in assessing pathophysiological factors particularly cerebral ischaemia where a close parallel has been established between primate research and the findings here reported in man. Evoked response recording is also of value in assessing the effects of treatment particularly induced hypertension and hypervolaemia (9). We have demonstrated it as of some use in prognosis either alone or in combination with other neurological and neurophysiological methods.

It is likely that such functional assessment will come to be the most significant technique of monitoring in cerebral ischaemia and its use in combination with the episodic techniques of assessment such as those of metabolism by E CAT scanning or blood flow by radioactive or non-radioactive Xenon, will provide further check correlations of the relationships between blood flow and function already established in the primate.

References

1. Anziska BJ, Cracco RQ (1981) Short latency SEPs to median nerve stimulation: comparison of recording methods and origin of components. Electroenceph Clin Neurophysiol 52:531-539

2. Branston NM, Symon L, Crockard HA, Pasztor E (1974) Relationship between the cortical evoked potential and local cortical blood flow following acute middle cerebral artery occlusion in the baboon. Exp Neurol 45:195-208

3. Brierley JB, Adams JH, Graham DI, Simpson JA (1971) Neocortical death from cardiac arrest. Lancet II:560-565

4. Greenberg RP, Stablein DM, Becker DP (1981) Non-invasive localisation of brainstem lesions in the cat with multimodality evoked potentials. J Neurosurg 54:740-750

5. Halliday AM, Wakefield GS (1963) Cerebral evoked potentials in patients with dissociated sensory loss. J Neurol Neurosurg Psychiat 26:211-219

6. Hume AL, Cant BR (1978) Conduction time in central somatosensory pathways in man. Electroenceph Clin Neurophysiol 45:361-375

7. Hume AL, Cant BR, Shaw NA (1979) Central somatosensory conduction time in comatose patients. Ann Neurol 5:379-384

8. Hunt WE, Hess RM (1968) Surgical risk as related to time of intervention in the repair of intracranial aneurysms. J Neurosurg 28:14-20

9. Kosnick EJ, Hunt WE (1976) Post operative hypertension in the management of patients with intracranial arterial aneurysms. J Neurosurg 45(2):148-154

10. Nakanishi T, Shimada Y, Sakuta M, Toyokura Y (1978) The initial positive component of the scalp recorded somatosensory evoked potential in normal subjects and in patients with neurological disorders. Electroenceph Clin Neurophysiol 45:26-34

11. Noel P, Desmedt JE (1975) Somatosensory cerebral evoked potentials after vascular lesions of the brainstem and diencephalon. Brain 98:113-128

12. Starr A, Achor LJ (1975) Auditory brain stem responses in neurological disease. Arch Neurol 32:761-768

13. Starr A, Hamilton AE (1976) Correlations between confirmed sites of neurological lesions and abnormalities of far-field auditory brainstem responses. Electroenceph Clin Neurophysiol 41:595-608

14. Stockard JJ, Rossiter VS (1977) Clinical and pathologic correlate of brain stem auditory response abnormalities. Neurology 27:316-325

15. Symon L, Hargadine J, Zawirski M, Branston NM (1979) Central conduction time as an index of ischaemia in subarachnoid haemorrhage. J Neurol Sci 44:95-103

16. Tsumoto T, Hirose N, Nonaka S, Takahashi M (1973) Cerebrovascular disease: changes in somatosensory evoked potentials associated with unilateral lesions. Electroenceph Clin Neurophysiol 35:463-473

17. Uziel A, Benezech J (1978) Auditory brainstem responses in comatose patients: relationship with brainstem reflexes and levels of coma. Electroenceph Clin Neurophysiol 45:515-524

18. Uziel A, Benezech J, Baldy Moulinier M, Duboin MP (1979) Etude des potentials evoques du tronc cerebral dans les comas traumatiques. Rev EEG Neurophysiol 9:202-206

19. Williamson PD, Goff WR, Allison T (1970) Somatosensory evoked potentials with unilateral cerebral lesions. Electroenceph Clin Neurophysiol 28:566-575

Clinical Trials of Brain Protection: Problems and Solutions

J. D. Miller

Department of Surgical Neurology, Western General Hospital, Edinburgh EH4 2XU, Scotland, Great Britain

Introduction

The concept of brain protection as a clinical therapeutic approach to patients with, or at risk from, ischaemic/hypoxic brain damage arises from three factors - the rapidity with which cerebral insults produce dysfunction, the difficulty in determining how much structural brain damage has occurred, and the irreversibility of changes in neuronal structure once they have occurred. Introduced at first in the form of barbiturate therapy as a protective regimen against regional cerebral ischaemia, brain protection now involves the use of several anaesthetic and sedative agents aimed at depressing the brain metabolically and thereby protecting it from both elevated intracranial pressure and damaging cerebral ischaemia. It is now applied in patients who have suffered acute cerebrovascular occlusions, cardiac arrest (7), subarachnoid haemorrhage (2), head injury (3) and global hypoxic insults, as in asphyxiation or drowning (6).

Well accepted as the concept may be, the proof that patients treated by such regimens do actually fare better than others treated on a more expectant basis has proved difficult to obtain. For the most part, these are critically ill patients, many of whom die or are disabled regardless of which current therapy is offered. Brain protection is not without its own risks and many factors need to be considered before an adequate trial of brain protection can be established. The need for such trials, however, is very great (5).

The mechanisms of brain protection

While the overall goal of brain protection is to produce an end result that is better for the patient than would have occurred without the regimen, it may not be clear how this desirable end result is to be achieved. Brain protection is usually aimed at reducing cerebral energy metabolism. This, in turn, reduces cerebral blood flow, and blood volume and thereby prevents or ameliorates intracranial hypertension. The reality of the situation may be quite different from this theoretical model, however. For example, a rapid reduction in intracranial pressure immediately following infusion of barbiturate may be due almost entirely to a reduction in arterial pressure so that the imagined benefit of reduced ICP is more than offset by the much greater insult of cerebral ischaemia. Some other agents used in brain protection are hypertonic; effects on intracranial pressure may in this case be due, not to any form of metabolic protection, but to the osmotic effects of the agent.

When an agent reduces the cerebral metabolic rate for oxygen, blood flow usually falls in a parallel manner. The patient is therefore not really being protected in the true sense because any further fall in cerebral blood flow, induced by an increase in ICP or fall in arterial pressure and uncompensated by autoregulation, will reduce blood flow below the level necessary to provide adequate tissue perfusion even in the metabolically reduced state. The possibly deleterious effect of inequalities between the supply of oxygen and glucose at a time when anaerobic metabolism is occurring also needs to be considered and brain protection regimens modified to take such possible factors into account.

Determining the benefits of brain protection

Because the therapeutic regimens associated with brain protection are not without their own risks, a clear benefit must be seen to result from the treatment if it is to be recommended on a regular basis. Anecdotal evidence is not acceptable and a key factor in the development of a successful clinical trial is a prudent selection of objectively verifiable end points (1).

One obvious end point is death versus survival. This may not be the satisfactory end point that it seems, however, because of the severity of the disorders with which advocacy of brain protective regimens is associated. For example, in severe head injury, it could be argued that in patients with overwhelming primary brain trauma no form of treatment can significantly influence the 90 percent mortality rate associated with the presence of fixed dilated pupils, absent eye movement reflexes, extensor rigidity and elevated intracranial pressure (4). Furthermore, in patients where the cerebral insult has occured as part of a severe systemic disease, such as myocardial ischaemia or hepatic failure, the original disease may itself cause death even though the cerebral complication is satisfactorily treated. While a reduction in mortality would be convincing evidence of efficacy of a brain protection regimen, the evidence is acceptable only if the treated and non-treated groups are adequately matched for the known factors influencing death versus survival.

The extent of morbidity or neurological deficit or dysfunction is much harder to classify objectively. What may be a trifling deficit to a farmer, may be devastating to a mathematician, and vice versa. The investigator may have to be satisfied with more limited end points, such as the frequency with which the protective agent reduced raised intracranial pressure, or the frequency with which raised intracranial pressure continues to require treatment after establishment of the protective regime.

Assessing the risks of brain protection

Patients in need of this form of treatment are already seriously ill. The addition of any further medical risk can detract greatly from the value of brain protection and deter the establishment of a proper clinical trial. Most data is available on patients treated with large doses of barbiturate, where the most immediate problem is production of undesirable abrupt arterial hypotension. This occurs mainly in under-hydrated patients and is a particular problem in young patients with head injury, in whom treating physicians often elect to produce a degree of under-hydration. Monitoring of central venous pressure is not a satisfactory guide and it is advisable in patients with severe head injury, particularly when this has been combined with other injuries, to monitor pulmonary arterial and pulmonary capillary wedge pressures using a Swan-Ganz catheter and to transfuse electrolyte solutions until normal pressures are attained. There is also the possibility that barbiturates increase the risk of severe lung infection in the comatose patient. Ciliary activity in the bronchial tree can be markedly reduced by barbiturates; this may form the basis for increased retention of secretions and a high incidence of pulmonary infection in comatose patients treated with barbiturate therapy.

In the author's experience of 40 patients treated in a randomised trial of barbiturate therapy following very severe head injury, there was some increase in the incidence of severe pulmonary infection in barbiturate treated patients but this was not statistically significant. In such patients it is, in any case, often difficult to determine whether patients develop adult respiratory distress syndrome as a result of the head injury, followed rapidly by infection, or whether the infection develops first to be followed by a clinical picture resembling ARDS.

The use of any newer drug therapy for brain protection in these severely injured and insulted patients will always be fraught with risks. Pilot experience with the agent in question, used in the appropriate patients, must be obtained before any large scale randomised trial is set up.

Knowing the outcome without brain protection

This is an essential prerequisite to considering any trial of brain protection regimens. It is only against the backdrop of secure knowledge of the mortality and severe morbidity of the untreated condition that the risks of therapy can properly be set and a randomised trial of therapy fairly presented.

It is first necessary to acquire a data bank of information on a substantial number of patients in whom the severity of the disease state has been prospectively and objectively described and in whom a fair assessment of the outcome has been made at defined intervals after the injury, illness or ictus. Clinical information is the most crucial, and must include age, sex, level of consciousness and details of the neurological examination, together with information on the cardiovascular and respiratory status of the patient. To this information may be added the level of intracranial pressure and severity of any intracranial hypertension, the presence of intracranial haemorrhage or vascular occlusion and other pertinent laboratory data. In gathering this data, a compromise has always to be made between creating too large a number of definitions so that each patient becomes unique, and creating so few subdivisions that crucial information is lost. Because of the nature of the problem, data must all be collected at an early stage, within hours of insult or injury. The data defining the severity of the disease state is correlated with the outcome, also objectively determined in a manageable number of categories. The smallest meaningful number of categories is the three-division collapse of the Glasgow Outcome Scale into good recovery and moderate disability, severe disability and vegetative and dead. Although the Glasgow Outcome Scale can be expanded from five into seven categories, this addition increases the number of patients needed for an adequate data bank. With even a three-point outcome scale, data banks of more than 200 patients are necessary before complete inter-series uniformity can be achieved.

Once this data has been collected, the range of mortalities in this group of patients can be measured and a realistic assessment made of the maximum possible benefit that successful brain protection could yield. The most important clinical factors influencing outcome can be established and checks made during any trial to ensure that the randomisation process is producing comparable incidences of these key factors in the control and treated groups. There are now several predictive algorithms by which the main clinical factors related to outcome can be not only enumerated but ranked in order of importance (8).

Clinical trials of brain protection

The information from the prior stages of the study can be used to plan a conclusive trial of the value of cerebral protection. The most appropriate categories of patient can be identified from the data bank (predicted mortality untreated more than 10% but less than 90%). An estimate should be made of the possible percentage improvement in outcome that might be possible, should cerebral protection be effective. Using the predictive algorithms, the number of patients necessary to detect this difference at an acceptable level of confidence (usually over 90 percent) is calculated. Collaboration with a bio-statistician is absolutely essential.

A double-blind randomised trial of control and treated patients is to be preferred. Because of the evident effects of the agents used in cerebral protection, however, blinding may not be possible. It is essential, however, that entry criteria are strict and that the trial be randomised so that patients with comparable degrees of severity of injury and disease are present in equal numbers in both series. It may be necessary to stratify the study to ensure that even with relatively low numbers in treated and control groups, the most important outcome predictor variables are evenly distributed. For head injury these would be age, level on the Glasgow Coma Scale with particular emphasis on motor response, signs of brain stem dysfunction, presence of an intracranial haematoma.

For other conditions, different outcome predictors will need to be identified. For every exclusion that is made, however, it becomes more and more difficult to obtain adequate numbers and a single centre is unlikely to have sufficient numbers of patients. Multicentre studies bring additional problems of data consistency and uniformity. Although one important erroneous conclusion that may come from a poorly designed and executed trial is that a beneficial effect for cerebral protection is claimed but does not actually exist, a more important error, as far as patients and their families are concerned, is to conclude that cerebral protection confers no benefit when such a benefit does truly exist.

References

1. Becker DP, Miller JD, Greenberg RP (1982) Prognosis after head injury. In: Youmans J. Ed. Neurological Surgery. 2nd edn. Saunders, Philadelphia, pp 2137-2174

2. Hoff JT, Marshall LF (1979) Barbiturates in neurosurgery. Clin Neurosurg 26:637-642

3. Marshall LF, Smith RW, Shapiro HM (1979) Acute and chronic barbiturate administration in the management of head injury. J Neurosurg 50:25-30

4. Miller JD, Butterworth JF, Gudeman Sk et al (1981) Further experience in the management of severe head injury. J Neurosurg 54:289-299

5. Miller JD (1979) Barbiturates and raised intracranial pressure. Ann Neurol 6:189-193

6. Rockoff M, Marshall LF, Shapiro HM (1979) High-dose barbiturate therapy in humans. Annal Neurol 6:194-199

7. Safar P (1980) Amelioration of post-ischemic brain damage with barbiturates. Stroke 15:1-9

8. Stablein DM, Miller JD, Choi SC, Becker DP (1980) Statistical methods for determining prognosis in severe head injury. Neurosurgery 6:243-248

Pharmacological Effects in Protective and Resuscitative Models of Brain Hypoxia

A. Wauquier, D. Ashton, C. Hermans, and G. Clinke

Department of Neuropharmacology Janssen Pharmaceutica, 2340 Beerse, Belgium

Introduction

The first candidate compounds for pharmacological protection of the brain were barbiturates, but interest began to wane after these compounds were shown to be less effective in models of global brain ischemia followed by reperfusion (14). The search is now on for better protective agents. Obviously, both simple tests of brain protection and complicated tests of brain resuscitation are required at different stages of this process. The present article discusses the usefulness and predictability of brain protection experiments giving examples from our work with etomidate and flunarizine. These results are then discussed in the light of what is known about these compounds in resuscitation experiments.

Protective agents

Although practically all the models of hypoxia in which protective effects of drugs are measured have been criticized as being "unclinical", they offer the great advantage of being quick enough to allow comparative quantitative pharmacological studies. Obviously, different types of models are required since in a new field it is not possible to decide on the most relevant model, since clinical studies are practically non-existent. In this situation, a purely heuristic approach must be followed. Since nearly all the models involve not only brain hypoxia but also to a varying degree respiratory problems, seizures, cardiovascular complications etc., activity in a large number of models is required in order to separate out compounds worth testing in supposedly more clinically orientated models. These models further serve for studying the structure-activity relationship of groups of compounds within and between tests.

Selection of substances

A large number of compounds have been tested in two basic screening models in rats: the histotoxic dysoxia test (potassium cyanide injection) and a hypoxic hypoxia test (100% nitrogen environment) (37,40). In the histotoxic dysoxia test, rats are given a lethal

Table 1 Survey of tests in which etomidate (as compared with barbiturates) was found to be beneficial

Type model test	Species	Parameters measured	Reference
Survival models			
- hypobaric hypoxia	mice	prolongation survival	Wauquier et al. 1981
- histotoxic dysoxia (potassium cyanide)	rats	survial	Wauquier et al. 1981
- hypoxic hypoxia (100% nitrogen)	rats	survival	Wauquier et al. 1981
- complete ischemia (decapitation)	mice	gasping: onset and duration	Wauquier et al. 1980
Functional models			
- asphyxia (hypercapnia)	rats	time to EEG silence and recovery	Wauquier et al. 1980
- bilateral carotid ligation	gerbils	neurological outcome, survival	Hermans et al. 1983
- hypoxic-ischemia (Levine preparation)	rats	neurological outcome, brain histology	Van Reempts et al. 1982
- histotoxic dysoxia	rats	reflexes, EEG, histology	Ashton et al. 1981
Resuscitation models			
- hypovolemia	rats	arterial BP, EEG, respiration, survival	Hermans et al. 1980
	dogs	arterial BP, EEG respiration, EEG, survival	Wauquier et al. 1980; 1981
- cardiac arrest	dogs	resuscibility, arterial BP, ECG, (EEG)	Mullie et al. 1981
- bilateral carotid occlusion and reperfusion	gerbils	neurological outcome, survival, brain histology	Hermans et al. 1982
Anticonvulsant tests			
- electroshock	mice	tonic and clonic convulsions, survival	Wauquier et al. 1980
- audiogenic seizures	mice	clonic and tonic convulsions, survival	unpublished
- max. metrazol	rats	tonic and clonic convulsions, survival	Desmedt et al. 1976
- s.c. metrazol	rats	clonic convulsions	Wauquier et al. 1980
- bicuculline	rats	tonic and clonic convulsions, survival	unpublished
- allylglycine	rats	tonic and clonic convulsions, seizures	Ashton and Wauquier 1979
- amygdaloid kindling	rats	clonic seizures	Ashton and Wauquier 1979

intravenous injection of 5 mg/kg potassium cyanide, whereas in the hypoxic hypoxia test, rats are brought in a 100% nitrogen environment for 1 min. In both tests, compounds are given before the hypoxic challenge and protection against mortality is the criterion of protection. When activity was found in these two tests, several of these substances were subsequently tested in a number of other models. From these results, it appeared that there are two groups of substances which were systematically found to possess antihypoxic properties: hypnotics, including barbiturates and etomidate, and substances known today as calcium entry blockers (41,42,43,44). These findings do not preclude that substances chemically unrelated to these classes of drugs, will be found.

Etomidate and flunarizine

That barbiturates were found active in various antihypoxic models is far from new. The studies with the non-barbiturate hypnotic etomidate were new and gave rise to a number of interesting observations differentiating etomidate from the barbiturates.

Table 1 lists the number of tests in which etomidate in comparison with barbiturates (mainly thiopental and pentobarbital) was found active. Note the fact that etomidate was found active in all the models that have been used, but even more important is the finding that protection occurred at hypnotic or at subhypnotic doses, and that its activity outlasts the duration of hypnosis (37,40). This would suggest that the antihypoxic property is relatively unrelated to hypnosis, though the l-isomer of etomidate which also lacks hypnotic property is devoid of protective effects. This further suggest that decreased metabolism is not sufficient to explain the protective effects, which is at variance with the idea that hypnotics have to be given at doses causing a flat EEG, supposedly associated with a maximal metabolic depression. At present, there is insufficient knowledge to interpret these findings adequately, but it is not excluded that some biochemical property (GABA-mimetic)

Table 2 Possible mechanisms involved in brain protection with etomidate

- reduction in $CMRO_2$	Renou et al. 1978
- reduction in CBF	Bidabe-Renou et al. 1978
- inverse steal effect	Bidabe-Renou et al. 1978
- decrease of the increased intracranial pressure	Schulte am Esch et al. 1978
- inhibition free fatty acid liberation	Nemoto et al. 1982
- immobilization	Cohn 1982
- suppression of seizures (and sedation)	Wauquier et al. 1979, 1980
- GABA-mimetic action	Ashton et al. 1981, 1983
- cellular protection by membrane stabilization?	

or a specific membrane effect (chloride ionophore) can both be related to hypnosis and protection (6). Table 2 lists possible mechanisms involved in brain protection achieved with etomidate.

In comparison with barbiturates, etomidate possesses a number of other properties some of which might be of cardinal importance for substances which are to be used clinically in brain protection. To these advantages as compared to barbiturates, belong the fact that etomidate has few cardio vascular effects and that in a hypovolemic situation, blood pressure increased to almost normal levels (38,39). In addition, etomidate has a large safety margin, especially in ventilated animals (19).

Table 3 lists the number of tests in which the Ca^{2+} entry blocker flunarizine in comparison with calcium antagonists was found active (1). The most important consideration here is that flunarizine lacks a number of properties traditionally considered to form the basis of brain protective effects, namely metabolic depression (Table 4). However, flunarizine is a potent 'calcium entry blocker', which refers to the prevention of intracellular calcium accumulation, postulated now by different authors, to be the 'final common pathway' of cellular intoxication (25,41), which offers a new approach worth further exploration.

Questions involved

The main question involved is whether activity found in the above tests has any predictive value with respect to so-called 'clinical models of hypoxia'. If one refers to models of resuscitation, the answer is difficult to predict and probably can only be resolved by testing the substances in such models. At this stage, it is only possible to say that compounds active in screening models have usually been shown to be active in resuscitation models. Further, whatever the model used, only clinical trials are able to give the appropriate answers.

It has, however, to be considered that brain protection is not only desired in cases where the hypoxic insults has already occurred, such as cardiac arrest, but also in cases where hypoxia is expected to occur. In this respect, it has already been shown that particular drugs, such as lidoflazine for example do have potent protective effects against ischemia of the heart (13). The present question is whether flunarizine has protective effects against brain hypoxia.

Brain resuscitation

Cardiac arrest models

Since much emphasis has been given to cardiac arrest models elsewhere in this symposium, this aspect will not be dealt with extensively here. In recent studies, two new findings are worth considering. The first deals with a study of Todd et al. (26) on cardiac arrest in cats

Table 3   Survey of tests in which flunarizine (as compared to calcium antagonists) was found to be beneficial

Type model test	Species	Parameters measured	Reference
Survival models			
- hypobaric hypoxia	chicken	crest color survival	Krstić et al. 1979
- histotoxic dysoxia (potassium cyanide)	rats	survial	Wauquier 1982 Wauqier et al. 1982
- hypoxic hypoxia (100% nitrogen)	rats	survival	Wauquier 1982 Wauquier et al. 1981, 1982
Functional models			
- asphyxia (hypercapnia)	rats	time to EEG silence and recovery	Wauquier et al. 1982
- passive avoidance memory test	rats	memory (retention)	Wauquier et al. 1982
- drinking behavior	rats	latency to drink and water consumption	Clincke and Wauquier 1982
- hypoxic-ischemia (Levine preparation)	rats	brain histology	Van Reempts et al. 1982
Cellular model			
- spreading depression	rats	depth of SD and recovery (EEG and DC potential)	Wauquier 1982 Wauquier et al. 1982
Resuscitation models			
- hypovolemia	rats	arterial BP, ECG, EEG, survival	unpublished
- cardiac arrest	rats	arterial BP, ECG respiration,	Hermans et al. 1982
- cardiac arrest	dogs	CBF	White et al. 1982
- bilateral carotid occlusion and reperfusion	gerbils	neurological outcome, survival	Hermans et al. 1982
Anticonvulsant tests			
- electroshock	mice	tonic convulsions, survival	unpublished
- audiogenic seizures	mice	clonic and tonic convulsions, survival	unpublished
- max. metrazol	rats	tonic convulsions, survival	Desmedt et al. 1976
- bicuculline	rats	tonic convulsions, survival	unpublished
- allylglycine	rats	tonic convulsions, survival	Ashton and Wauquier 1979
- amygdaloid kindling	rats	clonic convulsions	Ashton and Wauquier 1979
	dogs	clonic and tonic convulsions	Wauquier et al. 1979

Table 4 Possible mechanisms involved in brain protection with flunarizine

- antagonism of the impaired red blood cell deformability	De Clerck and David 1981
- prevention damage of endothelium	De Clerck 1982
- antagonism vasoconstriction (also of cerebral vessels) without inherent myogenic effects	Van Nueten et al. 1978, 1982
- improved cerebral blood flow	White et al. 1982, Wauquier et al. 1982
- suppression seizures	Wauquier et al. 1982
- prevention intracellular calcium overload	Van Reempts and Borgers 1982; Wauqier et al. 1981

where thiopental was used. It appears that some beneficial effect was found, though limited to a higher survival rate in the thiopental-treated cats. The most important finding in this study appeared to be the effects on the EEG. Todd et al. (26) described that in control cats an 'unusual post-resuscitation EEG pattern was found' and that thiopental suppressed this seizure pattern, which suggested that the improved survival rate might have been due to this effect. However, in this study thiopental did not improve neurological status. The reason for this remains unresolved, but it is not excluded that the cardio-respiratory depressant effects of thiopental, artificially antagonized by pressure increasing substances, are deleterious. In a recent study of Mullie et al. (21), it appeared that dogs resuscitated from cardiac arrest, which became 'dopamine-dependent' had a much lower neurological outcome score than dogs requiring less dopamine.

The second study attracting attention is that of White et al. (46), in which he found that flunarizine improved cerebral circulation in dogs resuscitated after 20 min complete circulatory arrest. He further found that lidoflazine-treated dogs had a better neurological outcome than control dogs. Much emphasis in these studies is given to the importance of high pressure values immediately during the reperfusion state and further the importance of continuous monitoring and intensive care for a sufficiently long time with a group of highly skilled personnel. One problem often arising in these studies, apart from the high cost, is that the 'highly skilled personnel' become even more proficient in resuscitation and post-resuscitative care during the long course of the experiment. This can lead to problems of a too great variability. Thus randomization and standardization become of paramount importance, but standardization is very difficult to achieve in the intensive care situation.

Ischemia by ligation of vessels

In recent experiments, we studied neurological outcome and survival following carotid ligation in gerbils anesthetized with ether, etomidate, thiopental or the combination of ether and etomidate or thiopental (17). The neurological scores and survival were better with etomidate than with ether and with both other drugs, ether anesthesia worsened the outcome. Again this study showed that etomidate was a better protective agent than thiopental, however, the experiment did not provide evidence that the drugs would improve resuscitation.

In further studies, we aimed to answer this question by giving drugs i.p. just before reperfusion after bilateral carotid occlusion. In one study, etomidate was compared with thiopental, studying neurological outcome, spontaneous behavior, EEG and brain histology (18). It was found that there was a significantly better neurological score after etomidate than after thiopental, both being better than the control-treated gerbils. Further, brain histology showed ischemic cell changes specifically in the hippocampus in the control animals, which was significantly improved following etomidate or thiopental treatment.

A similar study was carried out with flunarizine. Figure 1 shows the neurological outcome scores in control and flunarizine-treated gerbils. The scores used included an evaluation of corneal, pineal, auditive, righting and pain reflexes and spontaneous movements (17). As can be seen, there was a significantly better neurological outcome in the flunarizine-treated animals than in the control-treated gerbils.

Though there is no direct evidence in this experiment that the better outcome is due to an improved cerebral blood flow, the experiments by White et al. (46) and our experiments point in this direction. We measured cerebral blood flow, using a thermoflow measurement (46) in dogs in which a global incomplete ischemia was induced, by clamping the subclavian and brachiocephalicus for periods of 20 min with 1 h interval between the occlusion periods. In a first set of experiments, dogs were given an i.v. injection with 0.1 mg/kg flunarizine, 5 min before the first occlusion period. In control dogs, each period of occlusion resulted in a progressively more profound decrease in brain temperature. Flunarizine was able to significantly prevent this large decrease. The EEG of all seven control dogs failed to return after the 3rd period of occlusion, and this was followed by cardiovascular collapse. In contrast, in six out seven flunarizine-treated dogs, the EEG did not become isoelectric and the cardiovascular parameters appeared favorable after the 3rd period of occlusion. Though this method is a qualitative measure of tissue perfusion rather than a quantitative measure of CBF, the simultaneously measured EEG events point in the direction that the decreased brain temperature parallels the CBF. The reason for mortality is presently unclear, but is probably due to total cardiovascular collapse, which also appeared to be prevented by the flunarizine treatment.

That the beneficial effects of flunarizine are due to a vasodilating effect can be ruled out on the basis of two pieces of evidence. Firstly, flunarizine did not enhance cerebral blood flow in non-occluded dogs. Secondly, and this constitutes the major difference between

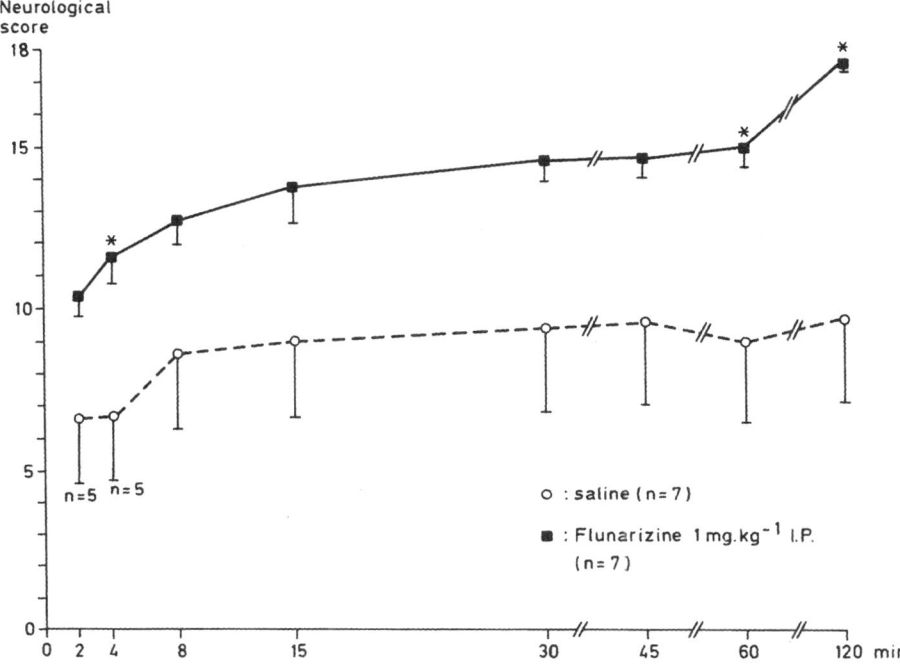

Fig. 1   Neurological scores obtained at different times (abscissa) after bilateral carotid occlusion for 5 min, followed by reperfusion in groups of 7 gerbils treated with saline or flunarizine given i.p. just preceding reperfusion. (Scores as defined in Hermans et al., 17). *p at least < 0.05.

flunarizine and other calcium antagonists: flunarizine is an antivasoconstrictive agent, without having inherent antimyogenic activity (31). Flunarizine inhibited the contractions of a variety of blood vessels, the most sensitive being the constriction of cerebral blood vessels, induced by Ca^{2+} (30,32,33). In adddition, it has been shown that flunarizine antagonized hypoxia-induced contractions of the basilar, middle cerebral and internal carotid arteries (34). It thus appears that flunarizine specifically antagonizes the vasospasms thereby improving tissue perfusion.

Conclusion

The present article demonstrates the advantages of protection models of hypoxia when studying the comparative pharmacology of brain protection. The evidence to date suggests that compounds active in a large number of protection models show some effects in resuscitation models. The experimental evidence gathered thus far in both protective and resuscitative models suggests that the advantages offered by etomidate, make it the drug of choice, in situations where sedation and/or immobilization are beneficial. The experiments with flunarizine suggest that calcium entry blockers form a new class of compounds, lacking some effects of earlier brain protective compounds, worthy of further development.

References

1. Amery W, Wauquier A, Van Nueten J, De Clerck F, Van Reempts J, Janssen PAJ (1981) The anti-migrainous pharmacology of flunarizine (R 14 950), a calcium antagonist. Drugs Exp Clin Res 7:1-10

2. Ashton D, Wauquier A (1979) Effects of some anti-epileptic, neuroleptic and gaba-minergic drugs on convulsions induced in rats by injection of D,L-allylglycine. Pharmacol Biochem Behav 11:221-226

3. Ashton D, Wauquier A (1979) Behavioral analysis of the effects of 15 anticonvulsants in the amygdaloid kindled rat. Psychopharmacol 65:7-13

4. Ashton D, Van Reempts J, Wauquier A (1981) Behavioral, electroencephalographic and histological study of the protective effect of etomidate against histotoxic anoxia produced by cyanide. Arch Int Pharmacodyn Thérap 254:196-213

5. Ashton D, Geerts R, Waterkeyn C, Leysen JE (1981) Etomidate stereospecifically stimulates forebrain, but not cerebellar, ^{3}H-diazepam binding. Life Sci 29:2631-2636

6. Ashton D, Fransen J, Wauquier A (1983) In vivo interactions between agonists and antagonists of the GABA-benzodiazepine-chloride ionophore receptor complex in a model of hypoxic hypoxia. In: Wauquier A, Borgers M, Amery W.Eds. Protection of tissues against hypoxia. Elsevier Biomedical Press, Amsterdam, in press

7. Bidabe-Renou AM, Constant Ph, Caille JM, Vernet J (1978) Vasoréactivité des infarctus cérébraux aux barbituriques et autres anesthésiques intraveineux. Ann Anesth Franc 19:821-826

8. Clincke GHC, Wauquier A (1983) Protective effects of flunarizine versus verapamil on water intake attenuated by normobaric hypoxia. In: Wauquier A, Borgers M, Amery W. Eds. Protection of tissues against hypoxia. Elsevier Biomedical Press, Amsterdam, in press

9. Cohn B Results of a feasibility trial to achieve total immobilization of patients in a neurosurgical intensive care unit with continuous etomidate infusion. Paper presented at "The First International Round Table on Intravenous Anesthesia" London (publication in preparation)

10. De Clerck F (1982) Effects of pharmacological agents on erythrocyte deformability and endothelial cell injury. J Cereb Blood Flow Metab 2 (Suppl 1):S50-S52

11. De Clerck F, David JL (1981) Pharmacological control of platelet and red blood cell function in the microcirculation. J Cardiovasc Pharmacol 3:1388-1412

12. Desmedt LKC, Niemegeers CJE, Lewi PJ, Janssen PAJ (1976) Antagonism of maximal metrazol seizures in rats and its relevance to an experimental classification of antiepileptic drugs. Arzneim Forsch 26:1592-1603

13. Flameng W, Xhonneux R, Borgers M (1983) Myocardial protection in open-heart surgery. In: Wauquier A, Borgers M, Amery W. Eds. Protection of tissues against hypoxia. Elsevier Biomedical Press, Amsterdam, in press

14. Gisvold SE, Safar P, Hendrickx HHL, Alexander H Thiopental treatment after global brain ischemia in pigtail monkeys (in preparation)

15. Hermans C, Wauquier A, Jageneau A, Francois Ph (1980) Etomidate vs. thiopental in haemorrhagic rats. 7th World Congr Anaesth, Hamburg, Abstracts, p 530

16. Hermans CFM, Fransen JF, Wauquier A (1983) Survival and neurological outcome after bilateral carotid ligation in the gerbil treated with ether, thiopental or etomidate. In: Wauquier A, Borgers M, Amery W. Eds. Protection of tissues against hypoxia. Elsevier Biomedical Press, Amsterdam, in press

17. Hermans CFM, De Reese R, Van Loon J, Loots W, Jageneau AHM (1982) A cardiac arrest model in rats for evaluating the antihypoxic action of flunarizine. Europ J Pharmacol 81:137-140

18. Hermans CFM, Van Reempts J, Wauquier A Neurological outcome, EEG and brain histology following post-treatment with etomidate and thiopental after bilateral carotid occlusion and reperfusion in the gerbil (in preparation)

19. Janssen PAJ, Niemegeers CJE, Marsboom RPH (1975) Etomidate, a potent non-barbiturate hypnotic. Intravenous etomidate in mice, rats, guinea pigs, rabbits and dogs. Arch Int Pharmacodyn Thérap 214:92-132

20. Krstić N, Radović A, Savovski K, Draganović D (1979) The influence of flunarizine upon the resistance to acute hypoxic hypoxia. Unpublished manuscript

21. Mullie A, Hermans C, Vandevelde K, Wauquier A (1981) Resuscitability of the heart with brain protective drugs during cardiopulmonary resuscitation in dogs. Critical Care Med 93:183

22. Nemoto E, Shiu GK, Nemmer JP, Bleyaert AL (1983) Pharmacologic attenuation of brain free fatty acid liberation during complete global ischemia as a measure of therapeutic efficacy. In: Wauquier A, Borgers M, Amery W. Eds. Protection of tissues against hypoxia. Elsevier Biomedical Press, Amsterdam, in press

23. Renou AM, Hunhilt J, Macroz P, Constant P, Billerey J, Khadarvo MY, Caille JM (1978) Cerebral blood flow and metabolism during etomidate anesthesia in man. Br J Anaesth 50:1047-1051

24. Schulte am Esch J, Pfeifer G, Thiemig F (1978) Der Einfluß von Etomidate und Thiopental auf den gesteigerten intracraniellen Druck. Anaesthesist 27:71-75

25. Siesjö BK (1981) Cell damage in the brain: a speculative synthesis. J Cereb Blood Flow Metabol 1:155-185

26. Todd MM, Chadwick HS, Shapiro HM, Dunlop BJ, Marshall LF, Dueck R (1982) The neurologic effects of thiopental therapy following experimental cardiac arrest in cats. Anesthesiology 57:76-86

27. Van Reempts J, Borgers M (1983) Morphological assessment of pharmacological brain protection. In: Wauquier A, Borgers M, Amery W. Eds. Protection of tissues against hypoxia. Elsevier Biomedical Press, Amsterdam, in press

28. Van Reempts J, Borgers M, Van Eyndhoven J, Hermans C (1982) The protective effects of etomidate in hypoxic-ischemic brain damage in the rat. Morphologic Assessment. Exp Neurol 76:181-195

29. Van Reempts J, Borgers M, van Dael L, Van Eyndhoven J, Van de Ven M (1982) Protection with flunarizine against hypoxic-ischemic damage of the rat cerebral cortex. A quantitative morphologic approach. Arch Int Pharmacodyn (in press)

30. Van Nueten JM (1978) Vasodilatation or inhibition of peripheral vasoconstriction? In: Vanhoutte P, Leusen I. Eds. Mechanisms of Vasodilatation. Karger, Basel, pp 137-143

31. Van Nueten JM, Van Beek J, Janssen PAJ (1978) Effect of flunarizine on calcium-induced responses of peripheral vascular smooth muscle. Arch Int Pharmacodyn Thérap 232:42-52

32. Van Nueten JM, Vanhoutte PM (1980) Improvement of tissue perfusion with inhibitors of calcium ion influx. Biochem Pharmacol 29:479-481

33. Van Nueten JM, De Clerck F (1982) Protection against hypoxia-induced decrease in tissue blood flow. In: Clifford Rose F, Amery WK. Eds. Hypoxia in the pathogenesis of migraine attacks. Pitman, London, pp 176-184

34. Van Nueten JM, De Ridder W, Van Beek J (1982) Hypoxia and spasms in the cerebral vasculature. J Cereb Blood Flow Metabol 2 (Suppl. 1):S29-S31

35. Wauquier A, Ashton D, Van der Starre P (1979) Anticonvulsant profile of etomidate, a non-barbiturate hypnotic. 11th Epilepsy Int Symp, Firenze, Astracts, p 135

36. Wauquier A, Ashton D, Melis W (1979) Behavioural analysis of amygdaloid kindling in Beagle dogs and the effects of clonazepam, diazepam, phenobarbital, diphenyl-hydantoin and flunarizine on seizure manifestation. Exp Neurol 64:579-586

37. Wauquier A, Ashton D, Clincke G, Niemegeers CJE, Janssen PAJ (1980) Etomidate: ein barbituratfreies Hypnotikum, antikonvulsive, antianoxische und hirnprotektive Wirkung in Tierexperiment. In: Opitz A, Degen R. Eds. Anästhesie bei zerebralen Krampf-anfällen und Intensivtherapie des Status epilepticus. Verlagsgesellschaft, Erlangen, pp 183-203

38. Wauquier A, Hermans C, Van den Broeck WAE, Jageneau A, Francois R (1980) Etomidate vs. thiopental and pentobarbital in haemorrhagic dogs. 7th World Congr Anaesth, Hamburg, Abstracts, p 295

39. Wauquier A, Van den Broeck WAE, Hermans C, Jageneau A, Francois P (1980) Electroencephalography in etomidate-, thiopental-, and pentobarbital-treated haemorrhagic dogs. 7th World Congr Anaesth, Hamburg, Abstract, p 433

40. Wauquier A, Ashton D, Clincke G, Niemegeers CJE (1981) Anti-hypoxic effects of etomidate, thiopental and methohexital. Arch Int Pharmacodyn Thérap 249:330-334

41. Wauquier A, Clincke G, Ashton D, Van Reempts J (1981) Considerations on models and treatment of brain hypoxia. In: Van Hof MW, Mohn S. Eds. Recovery from brain damage. Developments in Neuroscience. Vol. 13. Elsevier North Holland Biomedical Press, Amsterdam, pp 95-114

42. Wauquier A (1982) Brain protective properties of etomidate and flunarizine. J Cereb Blood Flow Metabol 2 (Suppl. 1):S53-S56

43. Wauquier A Effects of calcium entry blockers in models of brain hypoxia. In: Godfrain T, Herman A, Wellens D. Eds. Calcium entry blockers in vascular and cerebral dysfunctions. Westminister Publishers, New York, in press

44. Wauquier A, Ashton D, Clincke G, Van Reempts J (1982) Pharmacological protection against brain hypoxia: the efficacy of flunarizine, a calcium entry blocker. In: Amery WK, Clifford Rose F. Eds. Cerebral hypoxia in the pathogenesis of migraine attacks. Pitman, London, pp 139-154

45. Wauquier A, Melis W, Van Loon J The efficacy of flunarizine on CBF and EEG in a new model of global incomplete ischemia in dogs (in preparation)

46. White BC, Gadzinsky DS, Hoehner PJ, Krome PJ, Krome C, Hochner T, White J, Trombley JH (1982) Correction of canine cerebral cortical blood flow and vascular resistance after cardiac arrest using flunarizine, a calcium antagonist. Ann Emerg Med 11:118

The Clinical Use of Hypnotic Drugs in Head Injury

N. M. Dearden and D. G. McDowall

Department of Anaesthesia, 24 Hyde Terrace, Leeds, West Yorkshire LS2 9LN, Great Britain

Hypnotics such as thiopentone and althesin have been advocated for clinical situations in which brain ischaemia is anticipated or has occurred. In head injury, these drugs have been used with two separate philosophies, (i) that by reducing intracranial blood volume they assist in the control of intracranial pressure and (ii) by reducing brain oxidative metabolism, they provide some degree of protection against ischaemia. In Leeds, we use mainly althesin for this purpose, since, like thiopentone, it reduces cerebral blood flow (CBF), cerebral oxygen requirements and intracranial pressure (ICP) (2,4).

If the ICP is not controlled by an althesin infusion, given to a maximum rate of 40 ml/h in an adult, then a thiopentone infusion is substituted. Thiopentone is used ab initio if there is any history of allergy. The main advantage of althesin over the barbiturates is its rapid clearance, even after a prolonged infusion. The cerebral function monitor (cfm) (3) demonstrates a return of electrical activity within 3-6 h of stopping the infusion. The times to recovery of the cfm record after thiopentone are very much longer and are illustrated in Fig. 1.

In addition to infusions, althesin and thiopentone boluses are employed in head injury management to reduce the increases in ICP which occur with nursing procedures and chest physiotherapy (1). A further indication is the treatment of repetitive convulsive activity seen on the cfm. This is a less common indication and one in which etomidate may have a special role in view of its anti-convulsant action.

Our use of infusions of althesin and thiopentone to control ICP has increased during the past four years. In 76 severely head-injured patients treated up to July 1980, nine patients were so treated. In 102 similar head-injured patients admitted since August 1980, hypnotic drug infusion has been used in 32 patients. Although there is no doubt that these drugs can control ICP in a proportion of cases, the question still remains: Do they actually improve the outcome from severe head injury? The current presentation represents a preliminary attempt to answer this question.

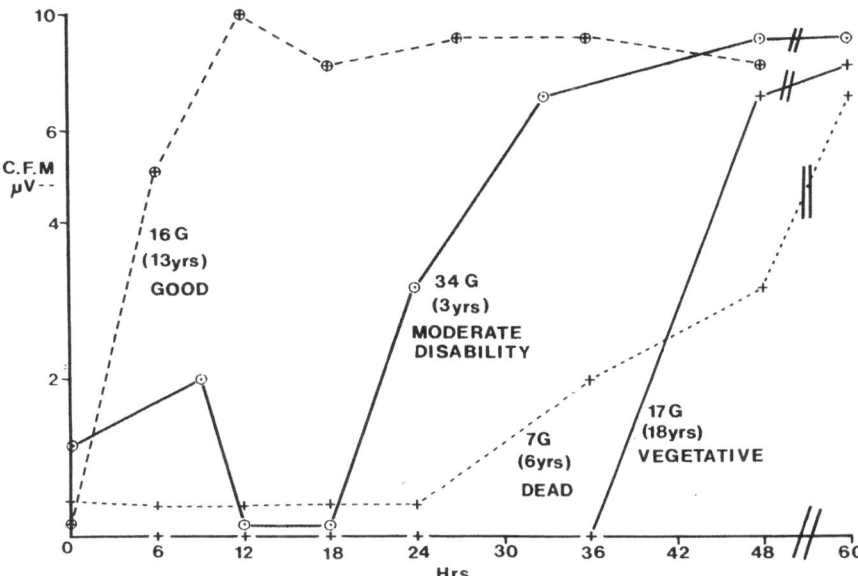

Fig. 1   The lower limit of the cerebral function monitor (cfm) trace is plotted against time since the discontinuance of thiopentone infusion in four head-injured patients. The total dose of thiopentone is given, together with the patient's age and outcome

Our indication for initiating an althesin or thiopentone infusion is a mean ICP of 25 mmHg maintained for at least 10 min or, alternatively, an ICP above 20 mmHg rising rapidly. Once the infusion is commenced, we attempt to control ICP to values below 25 mmHg.

39 of the 76 patients in the earlier series (when infusions were less commonly used) had ICP values above 20 mmHg and 25 of them died, giving a mortality of 64%. The outcome in the 32 patients treated recently for elevated ICP or "brain compression" on CAT scan is given in Table 1. From this it will be seen that, of the patients given infusions, 18 died and three were vegetative, giving a mortality of 60%, which is not significantly different from the mortality in the earlier series. Table 2 shows the outcome in relation to Glasgow Coma Score on admission to intensive care. It will be seen that there was no clear relationship between poor outcome and initial coma score. It seems that in this group of patients, in whom high ICP develops despite controlled ventilation and mannitol, the outcome does not relate closely to the initial clinical status.

In those patients whose ICP was not controlled by althesin, a barbiturate infusion was commenced. With the exception of one patient who received pentobarbitone, the infusion was 2.5% thiopentone. Of the 11 cases treated with barbiturates after ICP had escaped control by althesin, nine died or recovery was only to vegetative status. This suggests that

Table 1 Outcome and reason for hypnotic infusion (D/V in brackets)

	ICP=20 mm Hg and above	Brain compression	Sedation	Fits
Althesin only	12 (7)	2(1)	1(0)	1(0)
Althesin → barbiturate	11 (9)	0	0	0
Thiopentone only	3(3)	2(1)	0	0

30(21)
(70%)

Table 2 Outcome and initial coma scores (D/V in brackets)

Coma score	3 and 4	5, 6 and 7	8 and 8+
Althesin only	8(3)	5(3)	3(2)
Althesin → barbiturate	3(3)	7(5)	1(1)
Thiopentone only	1(1)	2(2)	2(1)

when althesin fails, the barbiturates are also usually ineffective. However, in the two cases who survived, the outcome was good in both.

It will be seen that there was a small group of patients who received thiopentone instead of althesin and, of these, only one of five survived. The group is too small to be worthy of further comment.

Our conclusions from these preliminary results are:
(i)   In patients undergoing intensive management following severe head injury, althesin infusions will often control ICP, and when they do so almost half the patients make a worthwhile recovery. When althesin fails, thiopentone usually is also ineffective, though occasionally a worthwhile recovery results.
(ii)   Despite the control of ICP achieved by these infusions, there is insufficient data available to indicate whether outcome is favourably affected. Because of the interplay of a number of factors in determining prognosis in severe head injury, more patients will need to be studied to define the influence of hypnotic drug infusions on outcome in sub-groups divided by age, coma score and presence of haematoma.

References

1. Moss E, Gibson JS, McDowall DG (1979) The effects of nitrous oxide, Althesin and thiopentone on ICP during physiotherapy in patients with severe head injuries. In: Shulman K, Marmarou A, Miller JD, Becker DP, Hochwald GM, Brock M. Eds. Intracranial Pressure IV. Springer, New York, pp 605-609

2. Pickerodt VWA, McDowall DG, Coroneos NJ, Keaney NP (1972) Effect of althesin on cerebral perfusion, cerebral metabolism and intracranial pressure in the anaesthetized baboon. Br J Anaesth 44:751-757

3. Prior P (1979) Monitoring Cerebral Function. Elsevier/North Holland, Amsterdam

4. Turner JM, Coroneos NJ, Gibson RM, Powell D, Ness MA, McDowall DG (1973) The effect of Althesin on intracranial pressure in man. Br J Anaesth 45:168-172

Indications for Cerebral Protection

D. Kling, W. Russ and G. Hempelmann

Abteilung für Anästhesiologie und Intensivmedizin, Justus Liebig-Universität, 6300 Giessen, FRG

A large number of patients who have survived cardiac arrest suffer from cerebral sequelae and eventually die because of this. More than 20% of the survivors present with permanent cerebral defects (1,15,25). These simple clinical facts will always initiate attempts to improve clinical outcome after cerebral ischemia. All brain protective measures have to aim at an optimal cerebral perfusion pressure, a low intracranial pressure and a normal arterial pressure (5,6).

Since more than 16 years cerebral protective effects of barbiturate therapy have been investigated among others by Goldstein (4), Michenfelder (11,12,13), Hoff (7), Smith (22), Safar (20), Nemoto (14), Fitch (3) and Sawata (21). Mechanisms by which barbiturate therapy might be effective have to be elucidated, although a reduction in brain metabolism has been generally accepted. Neurochemical, autoregulatory and neurogenic factors are additionally under discussion. Advantages of a high-dose barbiturate therapy have been reported by the above-mentioned investigators. Disadvantages are known as well, especially with respect to the cardiovascular system.

In 1980, Wauquier (24) reported that anti-hypoxic effects of hypnotics were most pronounced when they used etomidate, which has reduced brain metabolic rate, oxygen consumption and intracranial pressure similar to barbiturates. The reason for the mechanism is still unknown: it is discussed that etomidate causes membrane stabilization. The advantage of etomidate compared to barbiturates should be in a lesser depression of cardiovascular functions. Investigations by Lüben et al. (8) demonstrated that a rather high dose of etomidate (4 mg/kg body weight) does not have any advantage with respect to cardiovascular parameters.

A smaller dosage of 1 mg/kg body weight given within 15 minutes and followed by a continuous infusion of 1 mg/kg x h is sufficiently above the hypnotic level.

The above mentioned results encouraged us to add etomidate for brain protection in patients coming to cardiac surgery. Psychosis and delirious states do occur after open-heart surgery up to 30 or 60% (2,8,10). Disturbances like fear, restlessness, paranoid delusions, impaired orientation and changes in consciousness may be caused by many variables such as mean arterial pressure during extracorporeal perfusion, body temperature during bypass performance, nonpulsatile flow, duration of the extracorporeal circulation, hemodilution with consecutive metabolic disturbances, microembolism and other factors.

Methods and materials

506 patients have been included in our study, all of them being older than 18 years; patients with pre-existing psychic or neurologic deficits were excluded from this investigation. The dominating surgical procedure consisted of coronary bypass grafting (Fig. 1). Immediately after the beginning of extracorporeal circulation, a first group of 255 patients received 1 mg/kg body weight etomidate within 15 minutes, being followed by a continuous infusion of 1 mg/kg x h during the total bypass procedure. We used a highly concentrated solution of etomidate (125 mg/ml).

A control group of 251 patients did not receive any etomidate during bypass. All patients were selected according to the extracorporeal circulation journal number: patients with uneven numbers belonged to the etomidate group, patients with even numbers were selected as members of the control group. The management of extracorporeal circulation has been standardized with a pump flow of 2.3 l/min x m^2. The priming volume in all patients was the same, arterial pressure during extracorporeal circulation was kept between 50 - 100 mmHg with a perfusion pressure (P_{art}-P_{ven}) always above 40 mmHg. Arterial perfusion pressure was measured and recorded continuously by means of radial artery cannula and Statham transducer. In each patient, 5-minute values were summed up and thus a mean arterial pressure was calculated.

	Etomidate			Control		
	total	n	%	total	n	%
ACVB	177	49	28	175	39	22
AV	23	5	22	26	11	42
MV	20	9	45	23	6	26
AV + MV	3	3	100	4	2	50
ASD / VSD	13	2	15	6	-	-
ACVB + valve	15	8	53	9	5	55
Aneurysmectomy	2	1	50	5	-	-
others	2	-	-	3	-	-
	255	77	30	251	63	25

Fig. 1   List of the different surgical procedures in the etomidate group and the control group with regard to the appearance of psychic disturbances

All patients had been investigated one day before the operation and daily during the first five postoperative days; each time, psychic and neurologic states were judged. Psychic grading included lack of initiative, stuporous state, impaired orientation of time, location and personality, as well as restlessness. The neurologic check-up consisted of investigations of the cerebral nerves and of the tendon reflexes.

For comparison of the effect of hypothermia with the effect of etomidate, we performed an additional investigation: in 37 patients during extracorporeal circulation, visual evoked potentials (VEP) were recorded at different temperature levels. In 11 patients, VEP were recorded during steady state neuroleptanalgesia and after infusion of 1 mg/kg body weight x h etomidate. Basic considerations were described in recent publications (17,18) and by Reilly et al. (16).

Results

Mean duration of extracorporeal circulation was almost identical in both groups; the same applies to nasopharyngeal and rectal temperature during the hypothermic period. Mean arterial perfusion pressure - calculated in the above mentioned manner - was 66 mmHg in the etomidate group compared to 68.5 mmHg in the control group (Fig. 2).
Psychic disturbances were observed in 25% of the control group and in 30% of the etomidate group (Fig. 1). The highest incidence of psychic disturbances was seen in the most complicated surgical procedures such as double valve replacements, aorto-coronary bypass grafting plus valve replacement and aneurysmectomy (Fig. 1).

In both groups of patients, there was a fair number of patients with perioperative cardiopulmonary problems such as prolonged ventilatory support (>24 h) and/or prolonged catecholamine therapy (adrenaline $> 5\gamma$ /min for more than 3 hours); in the control group, we found 8% and in the etomidate group 20%. In both groups, psychic disturbances were significantly more often in patients with perioperative cardiopulmonary problems. Etomidate obviously did not influence the outcome significantly: there were 19% and 21%,

		Etomidate n = 255	Control n = 251
ECC	(min)	116,0 ± 40,7	114,4 ± 40,4
Nasopharyngeal temperature	(°C)	26,6 ± 1,3	26,3 ± 2,7
Rectal temperature	(°C)	29,1 ± 2,1	29,3 ± 2,0
Part during ECC	(mmHg)	66,0 ± 9,0	68,5 ± 10,0

Fig. 2   Mean ECC duration, nasopharyngeal and rectal temperature and mean arterial perfusion pressure during ECC

respectively, psychic disturbances in patients without cardiopulmonary problems; patients with perioperative cardiopulmonary problems presented with psychic disturbances in 86% of the etomidate group and in 83% of the control group (Fig. 3).

Patients with perioperative cardiopulmonary problems improved within the investigation period of 5 days almost at the same rate: 30% of the etomidate patients of this subgroup and 29% of the control patients of this subgroup. Those patients who improved within 5 days presented with psychic disturbances in 54% and 40% respectively (Fig. 4).

There were neurologic deficits after open-heart surgery in none of the patients without perioperative cardiopulmonary problems. Those with perioperative cardiopulmonary problems presented with neurologic deficits in 42% in the etomidate subgroup and in 22% in the control subgroup.

In the 37 patients in which visual evoked potentials were recorded at different temperature levels, the latencies of the major negative (n_2) and positive (p_2) components increased (Fig. 5) and the afterdischarge disappeared completely below $31^{o}C$. At $31^{o}C$, there was a 23% increase of n_2 latency as well as at $25^{o}C$ an 89% increase.

In the 11 patients in which VEPs were recorded during steady-state neuroleptanalgesia and after infusion of etomidate, the n_2 latency appeared in normothermic patients at 93.9 ms. Etomidate was causing a 9.6% increase in n_2 latency from 93.9 ms to 102.95 ms (Fig. 6) and a suppression of afterdischarge.

Discussion

Our results show that perioperative cardiopulmonary problems appeared twice as often in the etomidate treated group, which resulted in the higher percentage of neurologic deficits

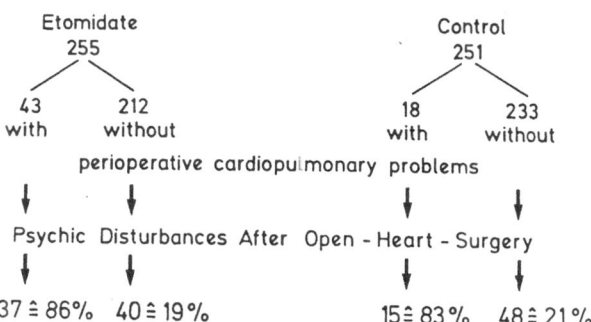

Fig. 3   Appearance of perioperative cardiopulmonary problems and psychic disturbances

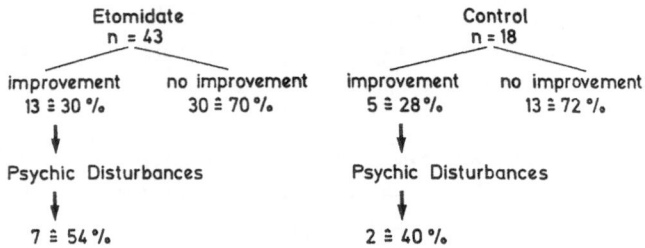

Fig. 4   Frequency of · improvement after perioperative cardiopulmonary problems connected with psychic disturbances

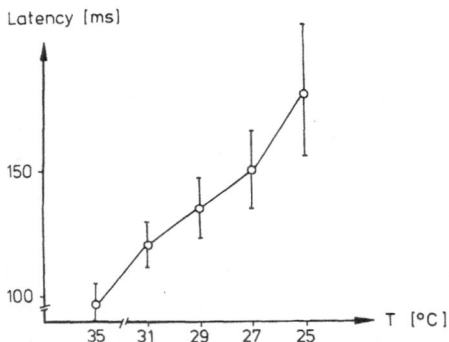

Fig. 5   Effect of hypothermia on mean latency of the major negative component of the visual evoked response (Russ et al., 19)

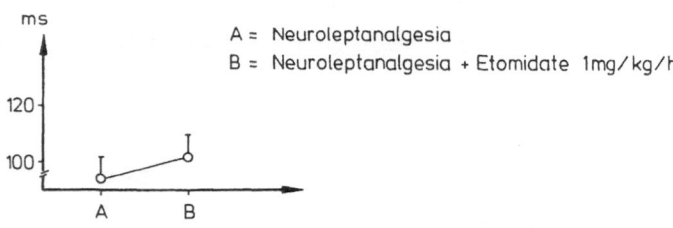

Fig. 6   Effect of etomidate on mean latency of the major negative component of the visual evoked potential

and psychic disturbances, although during bypass performance there was no significant difference of the mean arterial perfusion pressure between the two groups. At the present, there is no explanation for this higher incidence of cardiopulmonary problems. Mortality rates in both groups were the same (approximately 5%), which is not supporting results by Todd (23), pointing out that there has been a higher survival rate in an experimental group with barbiturate protection, neurologic deficits appearing more often, however.

When looking at the total groups and at the different surgical procedures, one can hardly state an advantage of etomidate with respect to psychic disturbances after open-heart surgery.

The results of the VEP investigation demonstrate that the effects of etomidate on visual evoked potentials are less - only 50% - than a temperature decrease to $31^{\circ}C$. This might add to our information on etomidate and its possible effects in cerebral protection.

In our opinion, there is no place for a prophylactic therapy with etomidate in open-heart surgery. It should be kept in mind that the effects of any hypnotic with respect to cerebral protection are less than with hypothermia. Avoidance of cardiopulmonary problems seems to be the most effective cerebral protective measure in cardiac surgery.

Etomidate might be administered in situations with incomplete cerebral ischemia, such as controlled deep hypotension, cerebral edema, tracheobronchial suction with increased intracranial pressure, total body hyperthermia, vasospasm and incomplete cerebral arterial stenosis. Total cerebral ischemia, global as well as regional are nevertheless indications for cerebral protective measures, e.g. addition of hypnotics such as etomidate. However, sound clinical proof is lacking so far.

Summary

The hypnotics thiopental and etomidate have been used for cerebral protection. 255 out of 506 patients undergoing open-heart surgery received etomidate during ECC for reduction of psychic disturbances after open-heart surgery.

Psychic disturbances were observed in 25% of the control group which did not receive etomidate and in 30% of the etomidate group. In both groups, there were patients with perioperative cardiopulmonary problems: 8% in the control group and 20% in the etomidate group. In these subgroups (patients with cardiopulmonary problems), psychic disturbances appeared more often (in percent) than in the total groups, 83% and 86% respectively. Patients improved after perioperative cardiopulmonary problems at the same rate: 30% in the etomidate subgroup and 28% in the control subgroup. However, they showed psychic disturbances in 54% and 40% respectively.

The investigation of visual evoked potentials demonstrated that the effect of etomidate on VEP is less than a fall in body temperature from 36°C to 31°C, which is adding to our information on cerebral protective measures. In our opinion, there is no place for a prophylactic cerebral protective therapy with hypnotics in open-heart surgery.

References

1. Bell JA, Hodgson JHF (1974) Coma after cardiac arrest. Brain 97:361-367

2. Blachly PH, Starr A (1964) Post-cardiotomy delirium. Am J Psychiatry 121:371-375

3. Fitch W (1981) Protection of the brain from ischaemia. Brit J Anaesth 53:201-202

4. Goldstein A, Wells BA, Keats AS (1966) Increased tolerance to cerebral anoxia by pentobarbital. Arch Int Pharmacodyn 166:138-143

5. Hempelmann G, Lüben V, Klug N (1982) Möglichkeiten der Hirnprotektion unter besonderer Berücksichtigung von Etomidat (Hypnomidate). Notfallmed 8:83-96

6. Hempelmann G, Kling D, Lüben V, v Bormann B (1982) Cerebral protection in neurosurgery, cardiac surgery and following cardiac arrest. J Cereb Blood Flow Metabol (Suppl 1) 2:66-71

7. Hoff JT, Smith AL, Hankinson HL, Nielson SL (1975) Barbiturate protection from cerebral infarction in primates. Stroke 6:28-33

8. Lüben V, Hempelmann G (1982) Hämodynamische Wirkung von hohen Dosen Etomidat zur cerebralen Protektion. Intensivmed Prax 4:67-74

9. Kornfeld DS, Zimburg S, Malm JR (1965) Psychiatric complications of open-heart surgery. N Engl J Med 243:287-292

10. Meyendorf R (1976) Psychische und neurologische Störungen bei Herzoperationen. Fortschr Med 94:315-321

11. Michenfelder JD (1982) Barbiturates for brain resuscitation: Yes and No. Anesthesiology 57:74-75

13. Michenfelder JD, Theye RA (1973) Cerebral protection by thiopental during hypoxia. Anesthesiology 39:510-517

12. Michenfelder JD, Milde HJ (1976) Cerebral protection by anesthetics during ischemia. Resuscitation 4:219-233

14. Nemoto EM (1982) Pathogenesis of cerebral ischemia-anoxia. Critical Care Med 6:203-214

15. Norris JR et al (1971) Anoxic brain damage after cardiac resuscitation. J Chron Dis 24:585-593

16. Reilly EL, Kondo C, Brunberg JA, Doty DB (1978) Visual evoked responses during hypothermia and prolonged circulatory arrest. Electroenceph Clin Neurophysiol 45:100-106

17. Russ V, Lüben V (1982) Der Einfluß von Etomidat in hypnotischer Dosis auf das visuelle evozierte Potential (VEP). Anaesthesist 31:483-484

18. Russ W, Lüben V, Hempelmann G (1982) Der Einfluß der Neuroleptanalgesie auf das visuelle evozierte Potential. Anaesthesist 31:575-578

19. Russ W, Hempelmann G Visual evoked potentials during extracorporeal circulation: Effect of hypothermia. Anesthesiology (submitted for publication)

20. Safar P, Bleyaert A, Nemoto EM, Moossy J, Snyder JV (1978) Resuscitation after global brain ischemia-anoxia. Critical Care Medicine 6:215-227

21. Sawata Y et al (1982) Acute tolerance to high-dose barbiturate treatment in patients with severe head injuries. Anesthesiology 56:53-54

22. Smith AL (1977) Barbiturate protection in cerebral hypoxia. Anesthesiology 47:285-293

23. Todd MM, Chadwick HS, Shapiro HM, Dunlop BJ, Marshall LF, Dueck R (1982) The neurologic effects of thiopental therapy following experimental cardiac arrest in cats. Anesthesiology 57:76-86

24. Wauquier A (1980) Hirnprotektive Effekte von Etomidat. Etomidate-Panel-Discussion: Weltkongreß für Anaesthesiologie

25. Willoughby JO, Leach BG (1974) Relation of neurological findings after cardiac arrest to outcome. Brit Med J 19:437-441

Prophylaxis of Cerebral Ischemic Damage From Vasospasm After Subarachnoid Hemorrhage

L. M. Auer

Universitätsklinik für Neurochirurgie, 8036 Graz, Austria

Introduction

Besides recurrent bleeding, cerebral vasospasm is a major complicating factor after subarachnoid hemorrhage (SAH) from cerebral aneurysm (20,21,33).

In the spontaneous course after SAH or in the interval between SAH and delayed surgery, symptomatic vasospasm is reported to occur in 20-55% of patients (19,30,34,35,37,42,55). Angiographically, "late" arterial spasm is mainly seen between day 5 and 21 after SAH in about 60% of patients (41,53); "early" spasm is less frequent, around 25%. In a series of 47 patients of our preoperatively graded I-III, 23% suffered deterioration from severe vasospasm (3).

The presence of an extensive subarachnoid clot was found to be highly correlated with vasospasm in several studies (16,39,40,48); 97% of patients with massive SAH on CT suffered neurological deficit combined with angiographical vasospasm.

Acute surgery, i.e. operation within 48-72 hours after SAH, has been advocated as the treatment of choice not only to prevent rebleeding, but to evacuate the subarachnoid blood accumulation, the source of substances causing arterial spasm (25,30,31,32,34,35,38,40,46).

The best results ever published with acute surgery were 16% (32) and 20% (30) incidence of symptomatic vasospasm postoperatively. No other treatment to overcome the problem was more successful until now, even though experimental studies have shown an unequivocal effect (2,9,17,18,43,44,45,52).

Although substances like prostaglandins (36), serotonin (1), hemoglobin (50,54), thromboxane A_2 (14,51) etc., have been made responsible for the development of arterial constriction likely to produce cerebral ischemia, the pathophysiological process is not yet fully understood - probably because of the large number of possible candidates, some potentiating the vasoconstrictor effect of others (5,51). All factors interact in allowing detrimental amounts of calcium-ions access into the vascular smooth muscle cells. Calcium is required for the contraction-relaxation and/or prevention of vasospasm. Substances that block calcium entry into the cell are therefore of interest.

The calcium antagonist nimodipine has been described as a drug to prevent cerebral ischemia from vasospasm, since it had shown the strongest predominantly cerebroarterial dilatatory effect when tested experimentally both in vitro (10,12,13,18) and in vivo (4,49).

Moreover, nimodipine increased cerebral blood flow in animal experiments (24,27) and in man (11,22). The antagonist action of nimodipine has been described to affect the "receptor-operated calcium channels", which are opened e.g. during stimulation with noradrenaline (29). To use this effect in clinical practice, the substance can be brought to the smooth muscle cells by perivascular or by intravascular administration of the drug. One approach had therefore been to inject nimodipine through a cisternal catheter in the postoperative period after acute aneurysm surgery in order to treat angiographically verified vasospasm in symptomatic patients. The first step to this approach had been intraoperative perivascular administration of nimodipine after clipping of the aneurysm to find out the lowest amount of the substance able to dilate the large vessels and resolve eventual spasm. A dosis of 200 μg had regularly been effective (6,8). Additionally, the postoperative incidence of symptomatic vasospasm seemed to be reduced and delayed following this single intraoperative treatment, suggesting a long-lasting protective effect of nimodipine.

Postoperative treatment of intracisternal injection of nimodipine turned out to be less effective than expected (5,6): neither vasodilation of normal vessel portions nor spasmolysis were achieved in all instances - probably due to some formation of scar tissue preventing free migration of the drug in the subarachnoid space surrounding the tip of the cisternal catheter. A further disadvantage of this approach became the necessity to withdraw the cisternal catheter 7 days after the operation in order to avoid the risk of infection. Patients developing symptoms after day 7 would thus not profit from such a procedure.

Vasospasm is known to extend from the large vessels at the skull base to the small branches in severe symptomatic cases. Therefore, as a second approach, the action of nimodipine on cerebral resistance vessels had been tested in patients during intravenous treatment (7).

Sixteen patients had been investigated during extra-intracranial arterial bypass operation, after craniotomy and opening of the dura and before performing the end-to-side anastomosis; in this randomized study, 1 μg kg^{-1} min^{-1} nimodipine or placebo was administered intravenously for a 10 min period (7). Pial vessel diameter variations were analyzed from photomicrograms. Pial arteries dilated less markedly compared to results from the animal study (4) although they dilate significantly by 8% after 5 min of nimodipine treatment (mean value from 66 arterial portions); 26 arteries smaller than 70 μm resting diameter dilated by

16%. In contrast to the data from cats, bigger arteries did not significantly dilate with $1 \mu g$ kg^{-1} min^{-1} after 10 min. The short observation period in patients may explain the lower degree of dilatation of small human compared to feline arteries. Mean systolic blood pressure in the nimodipine-treated group decreased by 18% (statistically insignificant) (7).

With this experience in mind (long-lasting dilatation of large vessels by intraoperative cisternal application and dilatation of resistance vessels by intravenous infusion), it seemed of interest to prevent rather than treat late symptomatic vasospasm and to develop a protocol for peri- and intravascular administration of nimopidine.

Present working hypothesis

Vasoconstrictor agents accumulate in the subarachnoid space leading to a slowly rising imbalance of normally existing dilator and constrictor substances and thereby cause the late type of vasospasm (51). Nimodipine, administered topically during surgery, binds loosely to the membranes of vascular smooth muscle cells in a high local concentration not achievable by systemic administration. The drug prevents excessive amounts of calcium ions from entering the cells via the receptor-operated channels and resolves pre-existing or intra-operative developing spasm. The nimodipine concentration required to balance the excess availability of constrictor agents in the subarachnoid space is maintained by continuous intravenous infusion of nimodipine. Infusion is started immediately after the topical administration and maintained for a period after which the constrictor agents have been cleared from the subarachnoid space in the majority of patients. Oral treatment with nimodipine for an additional period prevents spasm induced by sudden withdrawal of the drug in the rare circumstances of sufficient constrictor agents still present 2 weeks after SAH.

Intravenous infusion without initial topical administration may be insufficient to build up a high enough local concentration in order to counteract spasm developing pre-, intra- or immediately post-operatively.

Protocol of an open study for combined cisternal and intravenous administration of nimodipine

Patients of either sex between 15 and 70 years of age after a subarachnoid hemorrhage from a ruptured cerebral aneurysm may enter the study. The clinical status must conform to the Hunt & Hess grades I, II or III. Surgery must be performed within 72 hours after SAH in patients graded I or II, within 48 hours when graded III. Subarachnoid blood accumulation must be found on preoperative CT or at surgery. Preoperative four-vessel angiography is mandatory. In case of multiple aneurysms, treatment may be started only when the bleeding aneurysm(s) has (ve) been unequivocally identified and clipped. Cases with partial or total wrapping must be excluded. Likewise, patients with intracerebral hematoma, preoperative focal neurological deficit (e.g. patients with hemiparesis and/or aphasia) must be excluded in order to prevent postoperative difficulties in assessing further neurological deficit. Cranial nerve dysfunction alone does not necessarily exclude a patient; if a patient deteriorates in the immediate postoperative period due to surgery being complicated by premature rupture, massive bleeding, occlusion of important vessels, prolonged temporary clipping of major branches, the patient must be excluded from the series but may be treated and evaluated separately. Further exclusion criteria are pregnancy, renal or hepatic insufficiency, cardiac decompensation or severe arrhythmia.

CT is repeated on days 3 and 7 after the operation. Postoperative angiography should be performed when a patient deteriorates neurologically in the postoperative course in order to show or disclose sympathetic vasospasm.

Treatment protocol

During the operation, after aneurysm-clipping, a 2.4×10^{-5} M solution of nimodipine in Ringer's solution is applied to the vessels in the operating field in a volume of 20 ml to bath the vessels for a period of 10 min. Thereafter, intravenous infusion is started with 0.25 μg $kg^{-1}min^{-1}$ nimodipine and continued until day 14 after subarachnoid hemorrhage. From day 15 to 21, oral treatment with a daily dose of 4 x 60 mg nimodipine follows.

When symptomatic vasospasm is suspected postoperatively and verified angiographically, the infusion rate may be increased to 0.5 μg $kg^{-1}min^{-1}$.

Preliminary results

To date, 31 patients have entered the study; their age at the time of surgery ranged between 23 and 67 years. Angiograms revealed a single aneurysm in 29 patients, on the anterior communicating in 13, a middle cerebral in seven, the supraclinoidal portion of an internal carotid artery in six patients and of the basilar top in one. One patient had a ruptured pericallosal and an unruptured basilar aneurysm. Another patient had bilateral aneurysms of the internal carotid (the bleeding aneurysm could not be clearly identified; therefore both aneurysms were clipped in one session).

A short portion of vasospasm on preoperative angiograms, narrowing the lumen by less than 50% was seen in seven cases, of more than 50% in two. Severe spasm narrowing the lumen of internal carotid, middle and anterior cerebral arteries by more than 50% was found in one patient on day 3 after bleeding from an internal carotid aneurysm. Preoperative CT was not available for two patients; from the remaining 29, seven had extensive subarachnoid clots, 11 moderate and nine minor. Six patients were graded I, 12 patients II, and 13 patients III after Hunt and Hess.

The frontotemporal approach for access to the aneurysm was chosen in all but one case, the latter being operated on via the interhemispheric approach (25) for the pericallosal aneurysm.

Great care was taken to remove smoothly as much as possible of the subarachnoid blood accumulation without harming surrounding structures.

From the 25 patients in whom surgery itself caused no deteriorating complication, one patient experienced a short episode of speech disturbance and a flush TIA on the right arm on day 11 after surgery for a left middle cerebral aneurysm (i.e. day 13 after SAH). Angiography showed no vasospasm with the patient again being asymptomatic. Postoperative CT scans were normal throughout. The further postoperative course was uneventful; one month postoperatively, the patient had fully recovered and resocialized back to her previous work.

All other 18 patients took an uneventful course postoperatively with no signs of additional neurological deficit (except oculomotor palsy in the patient with basilar aneurysm). At three months followup investigation available from 27 patients, all were fully resocialized in the previous environment (Table 1).

Discussion

Acute surgery was chosen for the present study to prevent patients from rebleeding and in order to allow administration of a vasodilator agent, which was considered too dangerous preoperatively in patients with a ruptured aneurysm and the known high risk of rebleeding even without vasodilator treatment (3,20,33). Moreover, the evacuation of subarachnoid blood was considered important for the prevention of delayed symptomatic vasospasms.

Patients harboring intracerebral hematomas with surrounding tissue necrosis and edema preoperatively were excluded from this series, because experimental work has shown nimodipine to increase existing edema (23,47), the principal mechanism of action as well as the clinical significance of this observation still being unknown. Further studies will be required to elucidate the problem.

Subarachnoid blood accumulation in patients of this study as encountered on surgery was sufficient (severe and diffuse in 7, moderate in 7) to justify comparison of spasm-incidence with patients operated on acutely but not treated with nimodipine (30,32). Neither evacuation of subarachnoid blood during acute surgery nor absence of a subarachnoid clot preoperatively can thus account for results in the patients studied until now.

Regarding the effectiveness of nimodipine, a preliminary estimation from the limited material available from the present study shows that the biometrical prognosis is optimistic

Table 1

Status at 3 month follow-up	Preoperative grading (Hunt & Hess)		
	I	II	III
Patient can lead a full and independent life without neurological deficit	●●●●● ▲	●●●●● ●●●●	●●●●● ●▲
Patient can lead a full and independent life with minimal neurological deficit		▲	●● ▲ ▲

▲, complicated surgery; ●, uncomplicated surgery

Table 2 Prospective prognosis of patient-numbers required to show a significant effect of nimodipine, when assuming a 35%, 20% or 10% incidence of symptomatic vasospasm in the spontaneous course versus no case in the nimodipine-treated group (Fisher-Test)

Assumed incidence of vasospasm in the spontaneous course or after acute surgery	Level of significance for Nimodipine-effect	
	5%	1%
35%	5	11
20%	11	22
10%	80	69
	Nr. of patients required for study	

(see Table 2): 22 patients are required to provide significance at the 1% level when assuming a 20% incidence of symptomatic vasospasm after acute surgery without nimodipine treatment. 11 patients are required for the 5% level of significance (Fisher test).

Assuming an effect of nimodipine, the next question is, how does it really act?

Severe symptomatic spasm had occurred in two nimodipine-treated patients of a previous pilot series not only following topical treatment, but also after shorter intravenous infusion for 7 to 10 days after SAH and acute operation even during oral nimodipine treatment. Hemiplegia in one and complete motor aphasia in the other suddenly developed after two weeks; continuation of intravenous nimodipine reversed symptoms in both, although vasodilatation had not appeared on angiograms taken 10 to 30 minutes after start of infusion (3). Therefore, it was decided to develop a protocol with intravenous continuous drip for 14 days after SAH and acute surgery.

Data so far available do thus not fully support the working hypothesis:

Vasospasm detected in a second patient besides the one reported in the preliminary results of the ongoing study - though of moderate degree and after complicated surgery, still seems to develop to a certain extent even during nimodipine treatment. The favorable outcome of all patients suggests that the development of vasospasm of the large arteries is not fully suppressed but mitigated using the described protocol of nimodipine treatment; possibly by influencing the balance between physiological vasoconstrictor agents prostaglandin F_2 , serotonin etc. and the vasodilators such as prostaglandin D_2, E_2, G_2 and I_2 and prostacycline (15).

However, results from the previous study in patients who had been administered nimodipine intravenously during EC-IC-bypass surgery indicate the strongest effect of nimodipine on the small resistance vessels. Thus, preservation of adequate cerebral blood flow by lowering of resistance and opening of collaterals can be discussed.

Additionally, a second effect of nimodipine besides its vasodilatory action may be taken into consideration: thus, experimental cerebral ischemia was reduced with nimodipine (26,47) suggesting a protective effect via energy metabolism and calcium metabolism by preventing the activation of phospholipase and release of free fatty acids (47).

References

1. Allen GS, Gold LHA, Chon SN, French LA (1974) Cerebral arterial spasm Part 3: In vivo intracisternal production of spasm by serotonin and blood and its reversal by phenoxybenzamine. J Neurosurg 40:451-458

2. Anderson RGG (1972) cAMP and calcium ions in mechanical and metabolic response of smooth muscles: Influence of some hormones and drugs. Acta Physiol Scand (Suppl) 382

3. Auer LM, Gallhofer B, Ladurner G, Heppner F, Lechner H (1980) Spontanverlauf und operative Behandlung intrakranieller Aneurysmen nach Subarachnoidalblutung. Wien Med Wschr 24:871-875

4. Auer LM (1981) Pial arterial vasodilation by intravenous nimodipine in cats. Drug Res 31:1423-1425

5. Auer LM, Ito Z, Suzuki A, Ohta H (1981) Topical administration of nimodipine for treatment of cerebral vasospasm after acute aneurysm surgery. In: Proc Int Symp Cerebrovascular Diseases: New trends in surgical and medical aspects. Gardone Riviera, p 82

6. Auer LM, Ito Z, Suzuki A, Ohta H (1982) Prevention of symptomatic vasospasm by topically applied nimodipine. Acta Neurochir 63:297-302

7. Auer LM, Oberbauer RW, Schalk HV (1983) Human pial vascular reactions to intravenous nimodipine infusion during EC-IC bypass surgery. Stroke 14:210-213

8. Auer LM, Suzuki A, Yasui N (1982) Intraoperative topic nimodipine after aneurysm clipping. Neurochirugia (submitted for publication)

9. Bär HP (1974) Cyclic nucleotides and smooth muscle. Adv Cyclic Nucleotide Res 4:195-237

10. Brandt L, Andersson KE, Edvinsson L, Ljunggren B (1981) Effects of extracellular calcium and of calcium antagonists on the contractile responses of isolated human pial and mesenteric arteries. J Cereb Blood Flow Metabol 1:339-347

11. Brawanski A, Gaab MR, Bockhorn J, Haubitz I (1982) Atraumatic rCBF measurement: An aid in the timing of surgery and the management of spasm following SAH. Acta Neurochir 63:43-51

12. Edvinsson L, Andersson KE, Brandt L, Ljunggren B, MacKenzie ET, Skärby T, Young A (1981) Effects of Ca^{++} antagonists on cerebral blood vessels. J Cereb Blood Flow Metabol (Suppl 1) 1:S334-S335

13. Edvinsson L, Andersson KE, Brandt L, Högestätt E, Ljunggren B, Skärby T (1980) Studies of the effects of changes in extracellular calcium and potassium on cerebrovascular smooth muscle tone and of effects of various calcium antagonists on potassium contracted cerebral vessels. In: Betz E, Grote J, Heuser D, Wüllenweber R. Eds. Pathophysiol Pharmacotherapy Cerebrovasc Disorders. Witzstrock, Baden-Baden, pp 343-346

14. Ellis EF, Nies AS, Oates JA (1977) Cerebral arterial smooth muscle contraction by thromboxane A_2. Stroke 8:480-483

15. Ellis EF, Wei EP, Kontos HA (1979) Vasodilatation of cat cerebral arterioles by prostaglandins D_2, E_2, G_2 and I_2. Am J Physiol 237:H381-H385

16. Fisher CM, Kistler JP, Davis JM (1980) Relation of cerebral vasospasm to subarachnoid hemorrhage visualized by computerized tomographic scanning. Neurosurgery 6:1-9

17. Flamm ES, Yasargil MG, Ransohoff J (1972) Alteration of experimental cerebral vasospasm by adrenergic blockade. J Neurosurg 37:294-301

18. Flamm ES, Yasargil MG, Ransohoff J (1972) Control of cerebral vasospasm with parenteral phenoxybenzamine. Stroke 3:421-426

19. Flamm ES, Ransohoff J (1979) Subarachnoid hemorrhage and cerebral vasospasm. In: Pia HW, Langmaid CH, Zierski J. Eds. Cerebral Aneurysms. Springer, Berlin Heidelberg New York, p 152-155

20. French LA, Blade PS (1950) Subarachnoid haemorrhage and intracranial aneurysms. Lancet I:459-466

21. Gallhofer B, Auer LM (1982) Spontaneous course after subarachnoid haemorrhage - evaluation of 109 patients. Acta Neurochir 63:67-70

22. Gelmers HJ (1982) Effect of nimodipine (BAY e 9736) on postischaemic cerebrovascular reactivity, as revealed by measuring regional cerebral blood flow (rCBF). Acta Neurochir 63:283-290

23. Harris RJ, Branston NM, Symon L (1982) The effect of a calcium antagonist on the formation of cerebral ischaemic oedema and ion homeostasis. Proc 5th Int Symp Brain Edema. Groningen, p 56

24. Hoffmeister F, Kazda S, Krause HP (1979) Influence of nimodipine (BAY e 9736) on the postischaemic changes of brain function. Acta Neurol Scand 60, Suppl 72, 358-359

25. Ito Z (1982) The microsurgical anterior interhemispheric approach suitably applied to ruptured aneurysms of the anterior communicating artery in the acute stage. Acta Neurochir 63:85-99

26. Kazda S, Hoffmeister F (1979) Effect of some cerebral vasodilators on the post-ischaemic impaired cerebral reperfusion in cats. Arch Pharmacol Suppl 307, R43

27. Kazda S, Hoffmeister F (1980) Personal communication

28. Kazda S, Towart R (1980) Differences in the effect of the calcium antagonists nimodipine (BAY e 9736) and bencyclan on cerebral and peripheral vascular smooth muscle. Brit J Pharmacol 72:582P-583P

29. Kazda S, Towart R (1982) A new calcium antagonistic drug with a preferential cerebrovascular action. Acta Neurochir 63:259-265

30. Kikuchi H (1981) Emergency operation of ruptured aneurysm. In: Int Symp Cerebrovasc Diseases: New trends in surgical and medical aspects. Gardone Riviera

31. Ljunggren B, Brandt L, Kagström E, Sundbärg G (1981) Results of early operations for ruptured aneurysms. J Neurosurg 54:473-479

32. Ljunggren B, Brandt L (1982) The outcome of 100 consecutive cases of early aneurysm surgery. Acta Neurochir 63:215-219

33. McKissock W, Paine KWE (1959) Subarachnoid haemorrhage. Brain 82:356-366

34. Ohta H, Ito Z, Yasui N, Suzuki A (1982) Extensive evacuation of subarachnoid clot for prevention of vasospasm - effective or not? Acta Neurochir 63:111-116

35. Pasqualin A, Da Pian R (1982) An analysis of vasospasm following early surgery for intracranial aneurysms. Acta Neurochir 63:153-159

36. Pennink M, White RP, Crockarell JR (1972) Role of prostaglandin F_2 in the genesis of experimental cerebral vasospasm. Angiographic study in dogs. J Neurosurg 37:398-406

37. Perret G, Nishika H (1966) Cerebral angiography. J Neurosurg 25:98-116, 467-490

38. Saito I, Ueda Y, Sano K (1977) Significance of vasospasm in the treatment of ruptured intracranial aneurysms. J Neurosurg 47:412-429

39. Saito I, Shigeno T, Aritake K, Tanishima T, Sano K (1979) Vasospasm assessed by angiography and computerized tomography. J Neurosurg 51:466-475

40. Sano H, Kanno T, Sinomiya Y, Katada K, Katoh Y, Nakagawa T, Adachi K (1982) Prospection of chronic vasospasm by CT findings. Acta Neurochir 63:23-30

41. Sano K, Saito I (1978) Timing and indication of surgery for ruptured intracranial aneurysms with regard to cerebral vasospasm. Acta Neurochir 41:49-60

42. Sano K, Saito I (1979) Indication and timing of operation and vasospasm. In: Pia HW, Langmaid CH, Zierski J. Eds. Cerebral Aneurysms. Springer, Berlin Heidelberg New York, pp 208-216

43. Shepherd AP, Mao CC, Jacobson ED et al (1973) The role of cyclic AMP in mesenteric vasodilatation. Microvasc Res 6:332-341

44. Somlyo AP, Somlyo AV, Smiesko V (1972) Cyclic AMP and vascular smooth muscle. Adv Cyclic Nucleotide Res 1:175-194

45. Sundt TM, Onofrio BM, Merideth J (1973) Treatment of cerebral vasospasm from subarachnoid hemorrhage with isoproterenol and lidocaine hydrochloride. J Neurosurg 38:557-560

46. Suzuki J, Onuma T, Yoshimoto T (1979) Results of early operations on cerebral aneurysms. Surg Neurol 11:407-412

47. Symon L, Harris RJ, Branston NM (1982) Calcium ions and calcium antagonists in ischaemia. Acta Neurochir 63:267-275

48. Takemae T, Mizukami M, Kin H, Kawase T, Araki G (1978) Computed tomography of ruptured intracranial aneurysms in acute stage - relationship between vasospasm and high density on CT scan. Brain Nerve 30:861-866

49. Tanaka K, Gotoh F, Muramatsu F et al (1980) Effects of nimodipine (BAY e 9736) on cerebral circulation in cats. Drug Res 30:1494-1497

50. Toda N, Shimizu K, Ohta T (1980) Mechanism of cerebral arterial contraction induced by blood constituents. J Neurosurg 53:312-322

51. Towart R (1982) The pathophysiology of cerebral vasospasm, and pharmacological approaches to its management. Acta Neurochir 63:253-258

52. Triner L, Nahas GG, Vulliemoz Y et al (1971) Cyclic AMP and smooth muscle function. Ann NY Acad Sci 185:458-476

53. Weir B, Grace M, Hansen J, Rothberg C (1978) Time course of vasospasm in man. J Neurosurg 48:173-178

54. Wellum GR, Irvine TW, Zervas NR (1980) Dose responses of cerebral arteries of the dog, rabbit and man to human hemoglobin in vitro. J Neurosurg 53:486-490

55. Wilkins RH, Alexander JA, Odom GL (1968) Intracranial arterial spasm: A clinical analysis. J Neurosurg 29:121-134

Brain Protection by Barbiturate After Head Injury?
Clinical and Experimental Results

M. Sold, M. R. Gaab, B. Poch, and V. Heller

Abteilung für Anästhesie, Luitpoldkrankenhaus, Josef-Schneider-Strasse 2,
8700 Würzburg, FRG

Introduction

Barbiturates have been shown to exhibit a potentially beneficial effect in different models of cerebral ischemia, focal insult, and head trauma with raised intracranial pressure. Although the mechanism of action is not yet fully understood, it seems probable that "protection" is the consequence of the depression of functional metabolism of the nerve cells, thus not comparable to the true protection observed under hypothermia, and it is logical to assume that this unspecific effect may also be observed when using other drugs that equally depress $CMRO_2$ and CBF, such as althesin or etomidate. The subject of this article are clinical and experimental aspects of barbiturates in head trauma, especially their effect on intracranial pressure (ICP) and the development of brain edema.

Clinical effect on raised ICP

The ICP-reducing effect of barbiturates was first mentioned by Horsley in 1937 (8). Despite the work of Woringer in 1951 (21) and Søndergard in 1961 (19), systematic investigations did not begin until the early 70's and were undertaken by Shapiro's group (16,17). The use of barbiturates in the neurosurgical intensive care unit in cases with a critical increase in ICP was a consequence of their results and followed in 1977, a time in which monitoring of ICP became more and more a clinical routine and barbiturates enjoyed a general renaissance (10,11).

Figure 1 demonstrates the effect of barbiturates on raised intracranial pressure. Recorded were arterial pressure and ICP in a ten year old boy after head injury with ICP exhibiting typical plateau waves of the compensated type, as indicated by the fall of ICP below the pressure level the wave started from. Sufficient pentobarbital apparently had been given after the fourth dose; ICP then fell whereas arterial pressure remained constant. The CT scan of this patient showed hyperdensity rather than true edema, compatible with the diagnosis of "cerebrovascular engorgement".

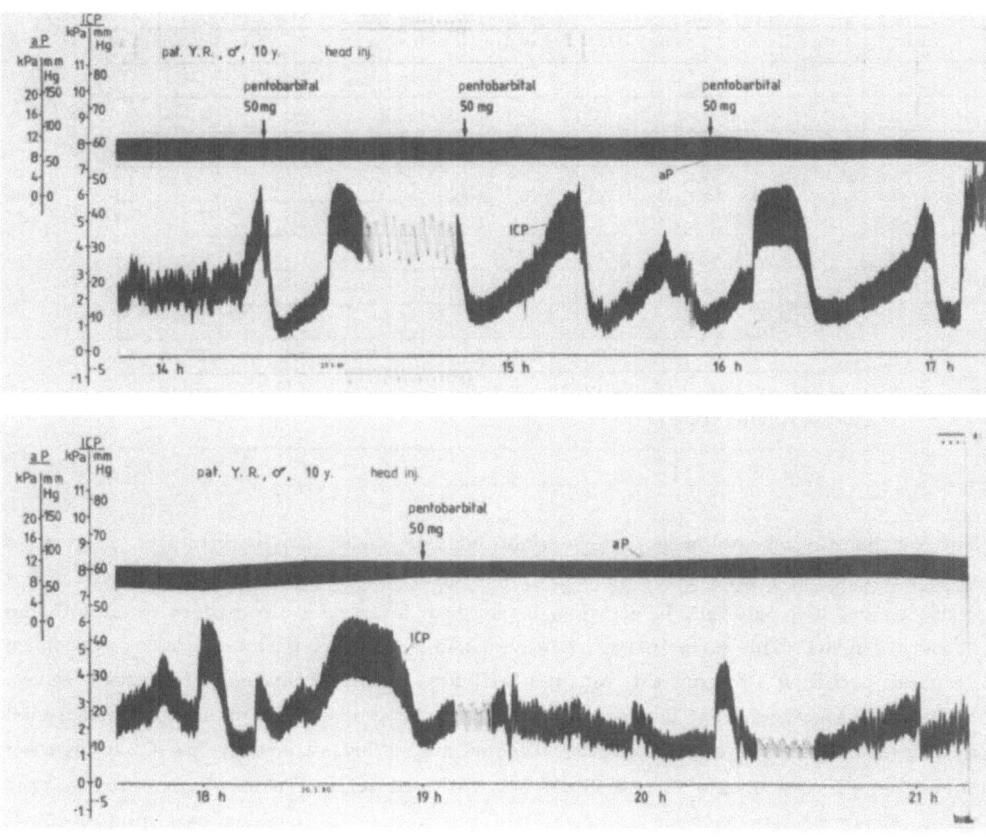

Fig. 1   Typical plateau waves in a 10 year old boy with closed head injury, pentobarbital
being effective after a cumulative dose of 200 mg

The reliability and efficacy of barbiturates in cutting off plateau waves could be
demonstrated in a large majority of more than 300 patients with head trauma that were
monitored for raised ICP (4). As is well known since the classical work done by Risberg and
his colleagues (15), plateau waves are accompanied by dilatation of cerebral arteries and, at
the same time, a decrease in cerebral blood flow that is caused by obstruction of the venous
outlet from the cranial cavity, probably due to compression of bridging veins (1). In this
situation, barbiturates are effective because they reduce cerebral blood volume. This is
illustrated in Fig. 2 which demonstrates the fall in ICP and the reduction in CBV after a
moderate dose of pentobarbital, determined in the areas of interest marked in the CT.
Although our method of measuring CBV (20) merely allows detecting relative changes, the
recent introduction of dynamic CT and emission computed tomography of ^{99m}Tc-labeled red
cells will soon add more information on alterations of cerebral blood volume in head trauma
(6,9).

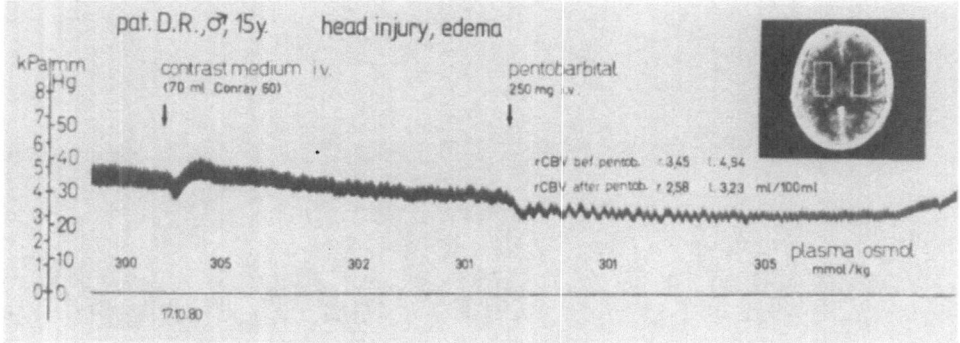

Fig. 2   Effect of pentobarbital (250 mg) on intracranial pressure (ICP) and cerebral blood volume (rCBV)

The mechanism of the reduction in blood volume caused by barbiturates suggests a metabolically mediated vasoconstriction, the result of the reduction in $CMRO_2$. As a consequence, ICP will fall. In addition, a decrease in arterial blood pressure adds to the reduction in ICP. This latter effect, however, can be dangerous since it may compromise cerebral perfusion. In contrast to their efficacy in the treatment of plateau waves, barbiturates have only a limited or no effect in progredient malignant intracranial hypertension when there is a hypodense edema suggesting an already low CBV. Figure 3 shows the ICP trace of a five year old boy with an isolated head trauma after a road accident. There was no bone fracture, and the initial CT revealed only mild swelling. However, on the second day ICP showed a progressive rise without plateau waves. Barbiturates were rather ineffective, and sorbitol only temporarily lowered ICP. A high amplitude/mean pressure ratio indicated vasoparalysis, and at an intracranial pressure of about 80 mmHg, bilateral mydriasis occurred. One can also see a Cushing reaction in this trace. A further example where barbiturates proved to be ineffective was the case of a 16 year old patient who suffered from traumatic occlusion of the left carotid artery. His CT showed a vast infarction of the left hemisphere with a huge collateral edema. He too, despite therapy, developed a progressive rise in ICP (Fig. 4). Bilateral mydriasis, which occurred when intracranial pressure approached systemic blood pressure, only initially responded to barbiturates and sorbitol. (In this patient, one can also see the transient increase in ICP that is the result of blood volume expansion caused by sorbitol.) The ICP trace that was recorded some hours later showed the persistent intracranial hypertension with small amplitudes and close coupling to systemic blood pressure that is characteristic for brain death. Thus, in the clinical situation, the barbiturates seem to be most effective when cerebral blood volume is increased or only slightly impaired. When a large edema predominates and CBV is already reduced, they probably do not influence ICP or outcome.

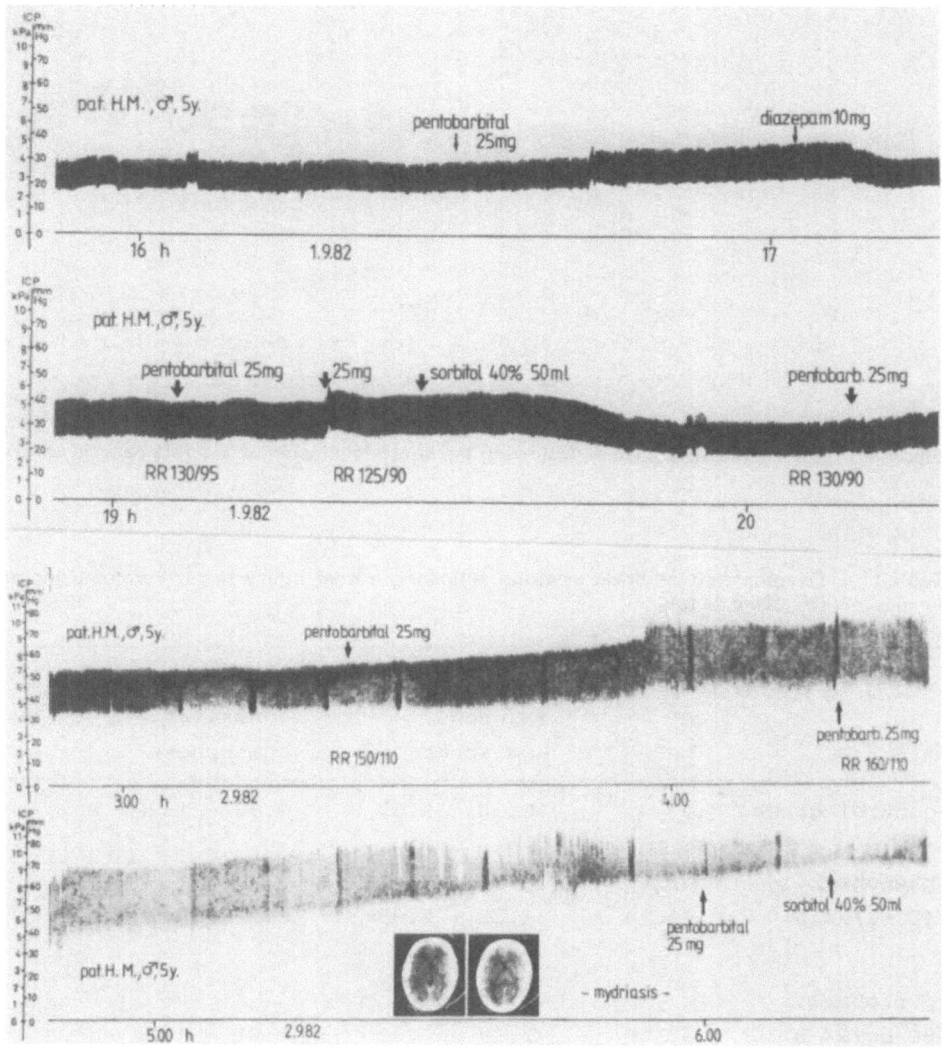

Fig. 3   Time course of ICP in a 5 year old boy with diffuse brain edema after head injury

Influence of barbiturates on experimental vasogenic edema

Besides the effect on ICP, barbiturates modify edema development. We could demonstrate this effect in a series of experiments carried out in rats. A standardized cold injury was inflicted on the intact skull over the right hemisphere. Therapy began 15 minutes after setting up the trauma and consisted of a continuous infusion containing pentobarbital, phenobarbital, or dexamethasone at varying doses. In this mode, ICP showed only a transient

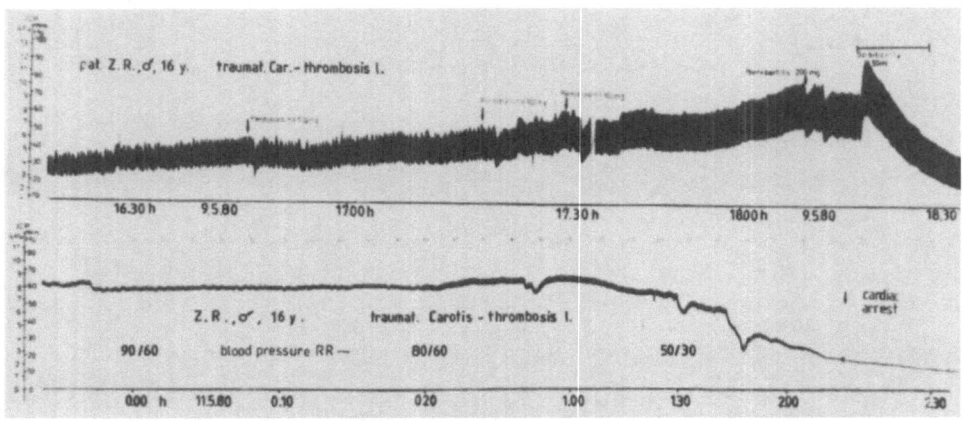

Fig. 4   ICP record of a patient following traumatic occlusion of the left carotid artery

Table I   Development of brain swelling following a cold injury in rats. Volume increase
(%) after 24 hours

| | n | VOLUME INCREASE (%) AFTER 24 HOURS | |
		injured hemisphere	contralat. hemisphere
control group	13	10.38 ± 2.57	0.13 ± 1.29
phenobarb. 40 mg/24 h	13	$4.30 \pm 1.98^*$	0.24 ± 0.79
pentobarb. 80 mg/24 h	13	$3.17 \pm 1.78^*$	0.28 ± 1.15
pentobarb. 120 mg/24 h	7	7.82 ± 4.13	0.56 ± 1.54

$\bar{x} \pm$ S.D.; $^*p < 0.01$ when compared to control

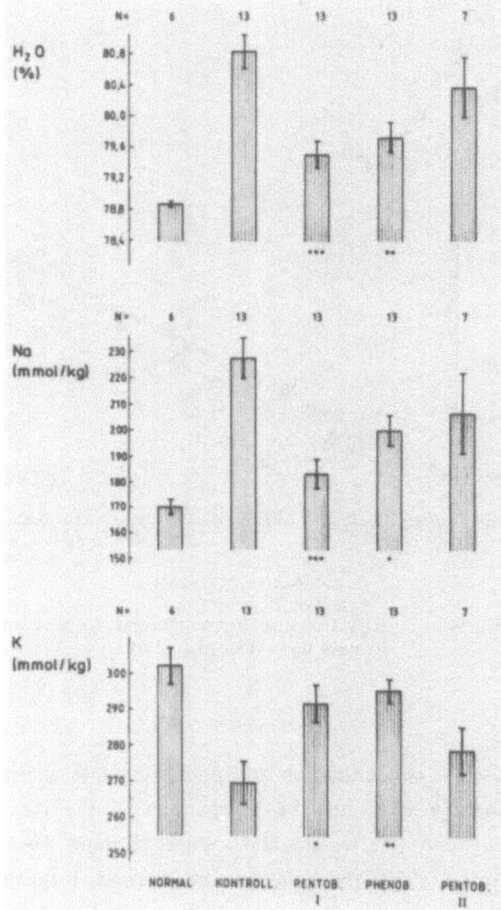

Fig. 5   Effect of an infusion of phenobarbital (40 mg/24 h) and pentobarbital (80 mg and
120 mg/24 h) on water content and sodium and potassium concentration of the
right hemisphere 24 hours after inflicting a cold injury in rats.
+ p < 0.05, +++ p < 0.005 when compared to controls

rise and then remained stable. In ventilated rats, barbiturate therapy resulted in an
increased survival rate when determined after 24 hours. In spontaneously breathing animals,
a positive effect of barbiturates was evident only after a dose of phenobarbital or
pentobarbital that did not compromise respiration or circulation (5,13).

In a further experiment, water content was determined 24 hours after the onset of trauma.
There was a smaller increase in the water content of the injured hemisphere when treated
with a moderate dose of pheno- or pentobarbital. The contralateral hemisphere showed no
development of edema (Table 1). The alterations in water content, together with those of

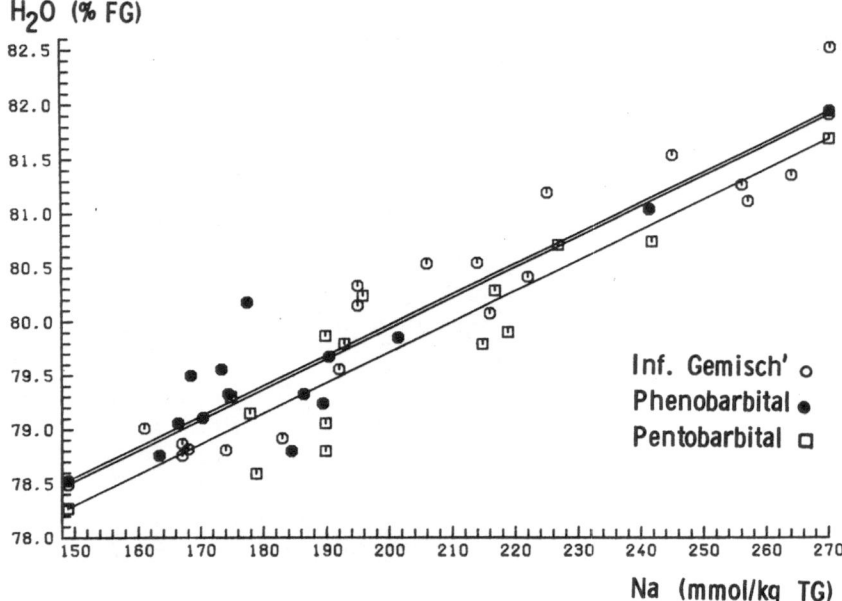

Fig. 6   Relationship between wet weight and sodium content of the injured hemisphere in rats with a cold lesion

sodium concentration and potassium loss again argue for a favorable effect of barbiturate therapy (Fig. 5). The comparison of the regression lines calculated for the relationship between wet weight and sodium content showed no significant difference between groups (Fig. 6). Thus the type of edema was not influenced. The pooled data yielded the equation

$$H_2O \ (\% \ ww) = 0.028 \times Na \ (mmol) + 74.2 \ (n = 45; \ r = 0.90).$$

The slope of the regression lines gained from these data did not differ from a theoretical regression that was calculated under the assumption that edema is an ultrafiltrate of plasma (2,7). Thus after cold brain injury, at least initially the resulting edema seems to be an ultrafiltrate of blood.

These observations on the effect of barbiturates in brain edema, corroborate the results published by Clasen et al. (3) and Smith and Marque (18). The reason for the lack of effect of barbiturates reported by Shapiro's group (17) may be the fact that they administered the barbiturates 24 hours after the trauma when edema already had developed. The mechanism by which barbiturates reduce the development of edema seems to be a decrease in the capillary transmural pressure gradient thus diminishing the "bulk flow" of fluid into the extracellular space (7,18). If an edema is already present and CBV not yet maximally reduced, barbiturates may precipitate a fall in ICP and thus reduce the danger of formation of intracellular edema caused by hypoxia. In addition, stress protection under barbiturate anesthesia may add to the favorable effect.

Metabolic effect of barbiturates in normal and in traumatized brain

To further study the changes caused by cold brain injury and the effects of barbiturates, multichannel surface microelectrodes were used to measure pO_2 and pH_2. Local tissue pO_2 was recorded and pO_2 histograms were calculated to give an estimate of tissue oxygen supply. In addition, after interrupting blood flow by inflating at tourniquet around the neck, the oxygen disapparance rate was determined as an indirect measure of oxygen consumption (20), i.e. the faster pO_2 falls down, the higher the oxygen consumption must have been. In addition, oxygen disappearance rates were correlated to the pO_2 measured just prior to the interruption of blood flow. To determine local blood flow, a constant hydrogen influx was established by adding hydrogen to the inhalation mixture. When local pH_2 had reached a plateau, hydrogen breathing was stopped. From the resulting hydrogen washout, local tissue blood flow was calculated (22).

The basic experiments were the same as before. A cold injury was produced in male Sprague-Dawley rats 24 hours before the actual measurements which were carried out in the area just outside the necrosis. For ethical reasons, we did no measurements in awake and immobilized animals, but used pentobarbital (65 mg/kg) and ketamine (320 mg/kg) for comparison. To be certain that the anesthetic given when inflicting injury did not influence the results, four groups of animals were chosen: pentobarbital - pentobarbital ketamine - ketamine, pentobarbital - ketamine, and ketamine -pentobarbital. However, since the choice of the first anesthetic proved to be irrelevant for the results obtained 24 hours later, data were combined to form two groups.

In healthy rat tissue, pO_2 histograms during anesthesia with pentobarbital and ketamine showed the typical distribution that is characteristic for an unimpaired tissue oxygenation with only a few low pO_2 values (Fig. 7). Both histograms were virtually identical. 24 hours after the cold injury, there was a significant shift to the right, a widening of the histogram, a loss of the normal distribution, but surprisingly there were less low pO_2 values than in healthy brain. Thus the histograms representative of edematous tissue showed a loss of the characteristic shape but did not prove an impairment of tissue oxygen supply. The reason for the rightward shift could be an uncoupling of blood flow and metabolism. When interrupting blood flow in healthy animals, the pO_2 disappearance rate was lower under pentobarbital as compared to ketamine. This is illustrated by the range of data points on the y-axis of the upper graph in Fig. 8 that is much smaller for the barbiturate than for ketamine; it is the simple consequence of the decrease in $CMRO_2$ caused by pentobarbital. When correlating the disappearance rate with the initial pO_2, a straight line was obtained. This means the higher the pO_2, the higher was the disappearance rate and therefore metabolism. One may

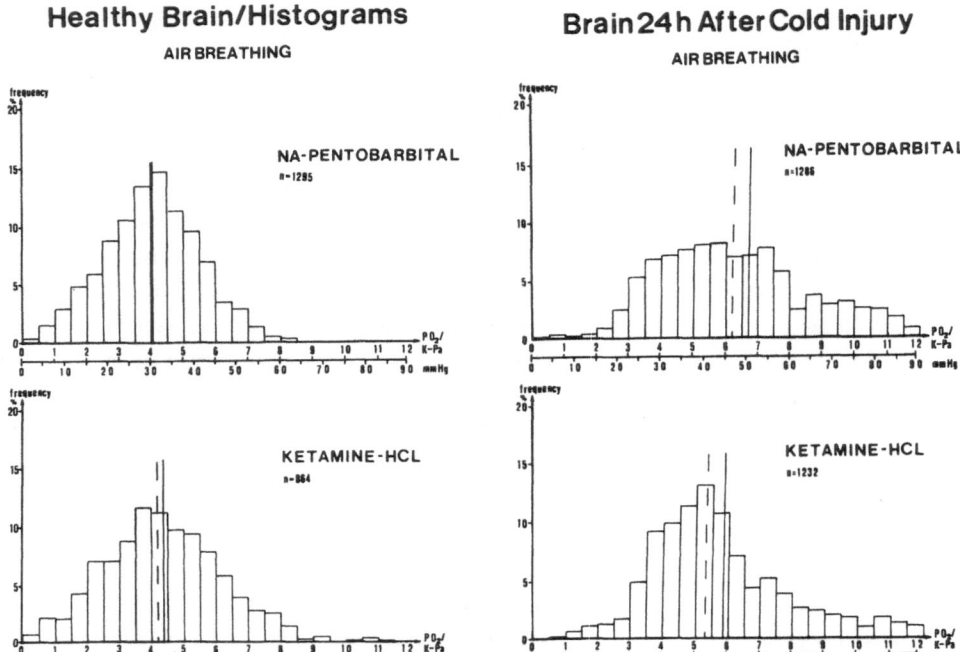

Fig. 7   pO$_2$ histograms in healthy brain and 24 hours after a cold injury in rats anaesthetized with pentobarbital or ketamine

conclude from this observation that autoregulation of blood flow and oxygen consumption was intact: a region where oxygen need was high indeed had a higher tissue pO$_2$. The slopes of both regression lines, however, were different indicating a higher tissue pO$_2$ for the same oxygen consumption under pentobarbital.

In injured brain, oxygen consumption again represented as the scatter of data points on the y-axis was lower, especially in rats anesthetized with ketamine (Fig. 8, upper right). Only when using pentobarbital could one detect disappearance rates that at least partially were comparable to normal. From this observation, one could argue in favor of the barbiturate. The regression lines were less steep when compared to healthy brain, and the correlation was worse, thus supporting the assumption of an impairment of autoregulation of regional flow and metabolism. The lower half of Fig. 8 shows the initial pO$_2$ plotted against the time

Fig. 8   Oxygen disappearance rates (dpO$_2$/dt) and times to zero oxygen pressure (t$_o$) vs. ➤ pO$_2$ at the moment of flow interruption in healthy brain and in rats subjected to cold brain injury. Left half: healthy brain, right half: injured brain

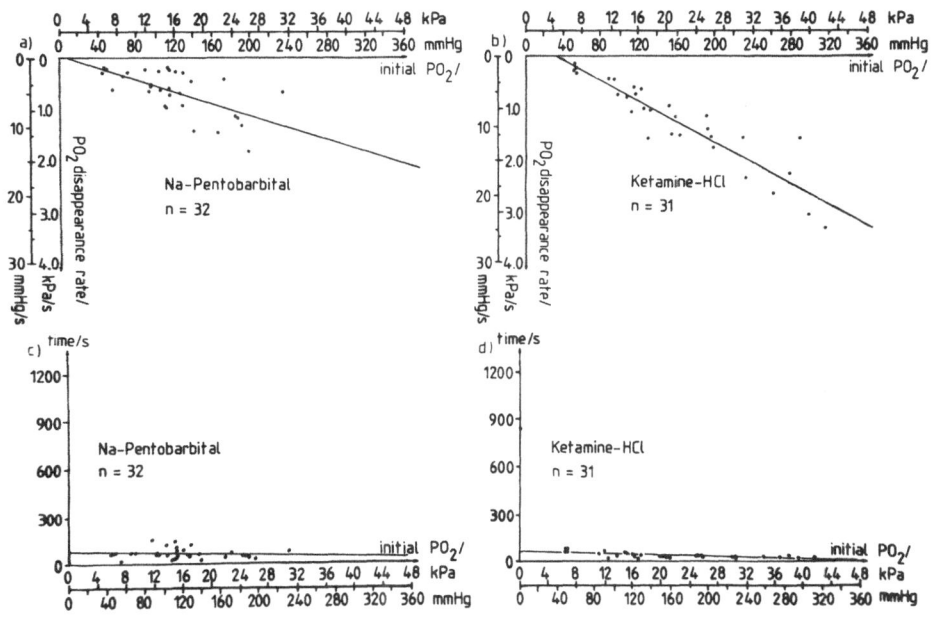

healthy brain

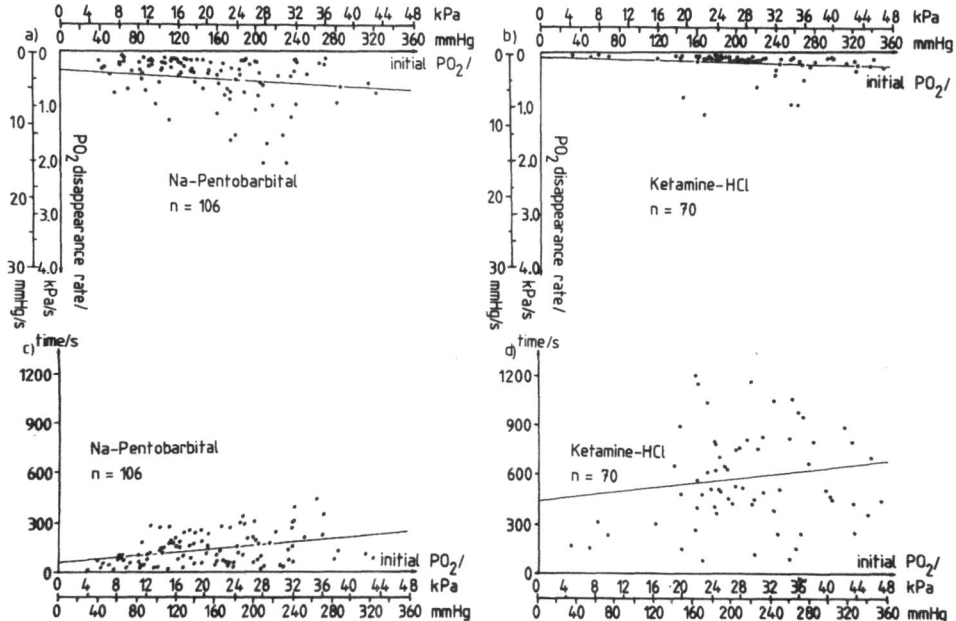

injured brain

that passed until tissue pO_2 fell to zero. In healthy brain, complete anoxia occurs very rapidly and, as indicated by the slope of the regression lines, the higher the pO_2 at the moment of blood stop, the quicker it falls. Thus a high pO_2 at the time blood flow is interrupted does not necessarily mean that the time until all oxygen has been consumed is prolonged. This again supports our hypothesis that regions with a high oxygen need are in fact supplied with a high amount of oxygen and that autoregulation is intact. In injured brain, this close coupling is interrupted, there is a large scatter of data (Fig. 8, lower right) and the slope of the regression line is altered. When comparing the anesthetics, one could argue in favor of the barbiturate that seemed to cause less disruption than ketamine. Unfortunately, in cold brain injury information on local blood flow and metabolism is scarce. However, just recently a fall in glucose utilization was reported in rats exposed to a small freezing lesion that fits well with our results (12).

In healthy animals, the measurement of mean local blood flow yielded a value of 50 ± 14 ml/100 g and min ($\bar{x} \pm$ SD) after pentobarbital and 109 ± 17 ml after ketamine. Whereas in traumatized rats mean local flow was not altered when the barbiturate was used (49 ± 30 ml), there was a significant reduction of flow in the ketamine group (37 ± 9 ml). This gives a clear contrast to the CBF values in normal tissue, where flow is higher under ketamine when compared to pentobarbital. From the flow measurements it is also obvious that the rightward shift of the pO_2 histograms that occurred in traumatized brain must not be interpreted as a consequence of reactive hyperemia. As indicated by the lower oxygen disappearance rates in injured brain, the rightward shift is probably caused by a decrease in oxygen consumption. The reason why the oxygen that is delivered to the tissue is not adequately utilized in the cells is not clear. Further experimental studies are required, and a definite answer will be obtained only when our merely physical methods are combined with biochemical and audioradiographic measurements.

References

1. Arnold H, Laas R (1980) Plateau waves: production in the rat and simulation by means of a mechanical model. In: Shulman K, Marmarou A, Miller JD, Becker DP, Hochwald GM, Brock M. Eds. Intracranial pressure IV. Springer, Berlin Heidelberg New York, pp 525-529

2. Baethmann A, Schmiedek O (1973) Pathophysiology of cerebral edema: Chemical aspects. Advanc Neurosurg 1:5-18

3. Clasen RA, Pandolfi S, Casey D jr (1974) Furosemide and pentobarbital in cryogenic cerebral injury and edema. Neurology 24:642-648

4. Gaab M (1980) Die Registrierung des intrakraniellen Druckes. Grundlagen, Techniken, Ergebnisse und Möglichkeiten. Thesis for habilitation, Würzburg

5. Gaab M, Herrmann F, Kerscher J, Rausch K, Lochner J, Pflughaupt KW (1979) Comparison on the effects of dexamethasone, barbiturate and THAM on experimental brain edema. Acta Neurochir 493, Suppl 28, pp 493-497

6. Hacker H (1982) Dynamic CT in cerebrovascular pathology. In: Cecchini A, Nappi G, Arrigo A. Eds. Neuroradiological and neurophysiological conditions. Emiras, Pavia, pp 223-224

7. Harbaugh RD, James HE, Marshall LF, Shapiro HM, Laurin R (1979) Acute therapeutic modalities for experimental vasogenic edema. Neurosurg 5:656-665

8. Horsley J (1937) The intracranial pressure during barbital narcosis. Lancet 232:141-143

9. Kuhl DE, Alavi A, Hoffman EJ, Phelps ME, Zimmerman RA, Obrist WD, Bruce DA, Greenberg JH, Uzzell B (1980) Local cerebral blood volume in head-injured patients. Determination by emission computed tomography of ^{99m}Tc-labeled red cells. J Neurosurg 52:309-320

10. Marshall LF, Shapiro HM (1977) Barbiturate control of intracranial hypertension in head injury and other conditions: iatrogenic coma. Acta Neurol Scand 56, Suppl 64, pp 156-157

11. Marshall LF, Smith RW, Shapiro HM (1979) The outcome with aggressive treatment in severe head injuries. Part II: Acute and chronic barbiturate administration in the management of head injury. J Neurosurg 50:26-30

12. Pappius HM, Wolfe LS (1982) Effect of drugs on local cerebral glucose utilization in traumatized brain: mechanism of action of steroids revisited. In: 5th Int Symp Brain Edema, Groningen, Abstr 24, p 44

13. Rausch A (1982) Barbiturate in der Behandlung des traumatischen Hirnödems. Eine tierexperimentelle Untersuchung. Inaugural dissertation, Würzburg

14. Reneau DD, Halsey JH (1978) Interpretation of oxygen disappearance rates in brain cortex following total ischaemia. In: Silver IA, Erecińska M, Bicher HI. Eds. Oxygen transport to tissue. Vol III. Plenum Press, New York, pp 189-198

15. Risberg J, Lundberg N, Ingvar DH (1969) Regional cerebral blood volume during acute transient rises of the intracranial pressure (plateau waves). J Neurosurg 31:303-310

16. Shapiro HM, Galindo A, Wyte SR, Harris AB (1972) Acute intracranial hypertension during anesthetic induction: partial control with thiopental. Europ Neurol 8:118-121

17. Shapiro HM, Galindo A, Wyte SR, Harris AB (1973) Rapid intraoperative reduction of intracranial pressure with thiopentone. Brit J Anaesth 45:1057-1062

18. Smith AL, Marque JJ (1976) Anesthetics and cerebral edema. Anesthesiology 45:64-72

19. Søndergard W (1961) Intracranial pressure during general anaesthesia. Dan Med Bull 8:18-25

20. Wodarz R, Gaab M, Pflughaupt KW, Nadjmi M (1980) Measuring regional cerebral blood volume (rCBV) by routine CAT scanning. Methods and results. In: Advanc Neurosurg. Vol 9. Springer, Berlin Heidelberg New York, pp 374-377

21. Woringer E, Brogly G, Schneider J (1951) Étude de l'action des anesthésiques généraux usuels sur la pression du liquide céphalo-rachidien; remarques pratiques quant à leur utilisation en chirurgie nerveuse. Anésth Analg (Paris) 8:649-662

22. Young W (1980) H_2 clearance measurement of blood flow: a review of technique and polarographic principles. Stroke 11:552-564

Thiopentone in the Treatment of Severe Head Injury: Is Raised Intracranial Pressure the Sole Indication for Its Use?

K. Wiedemann

Abteilung für Anästhesie, Universität Heidelberg, Im Neuenheimer Feld 110, 6900 Heidelberg, FRG

Prompted by the extensive experience of Bruce et al. (4) and Marshall et al. (14), we induced barbiturate coma in 27 brain trauma patients with intracranial pressure (ICP) raised to more than 25 torr for more than 15 minutes, and unresponsive to dexamethasone, mannitol, furosemide and moderate hyperventilation, as in a boy exhibiting Lundberg waves and pronounced ICP reactions to nursing (Fig. 1).

Methods and Patients

Special monitoring comprised intraarterial and intracranial pressure, one-channel bipolar temporal EEG and in most cases thiopentone plasma analysis by gas chromatography, twice daily. The epidural pressure transducer (9) requiring 4 mm thickness of skull bone for proper placement, was unsuitable in two children of 3 and 5, in whom raised ICP was assumed on CT scan and clinical symptoms.

The clinical features of the 13 survivors and 14 nonsurvivors are shown in Table 1. Nonsurvivors were significantly older than survivors (Wilcoxon test, $\alpha \leq 0.05$), their coma scores were slightly lower. However, all but one surviving patient scored between 3 and 7 during the first 24 hours following admittance.

Incidence of intracerebral hematoma was distributed almost equally, as was the case with contusions. Diffuse swelling occurred twice in non-survivors, but four times in survivors. In non-survivors, pupillary abnormalities were seen more often, and bilaterally fixed pupils occurred twice. Convulsions were seen once in each group.

ICP monitoring was commenced 24 to 48 hours following admittance, thiopentone infusion followed as indicated by intracranial pressure readings. With additional therapy maintained, thiopentone was administered by rapid infusion of 10 - 15 mg/kg during 5 to 10 minutes, taking care not to compromise cerebral perfusion pressure. Constant rate infusion was continued with 4 - 5 mg/kg . h, but sometimes had to be increased to 6 to 8 mg/kg . h to control ICP in the range of 20 to 25 torr and to achieve a burst-suppression pattern in the EEG.

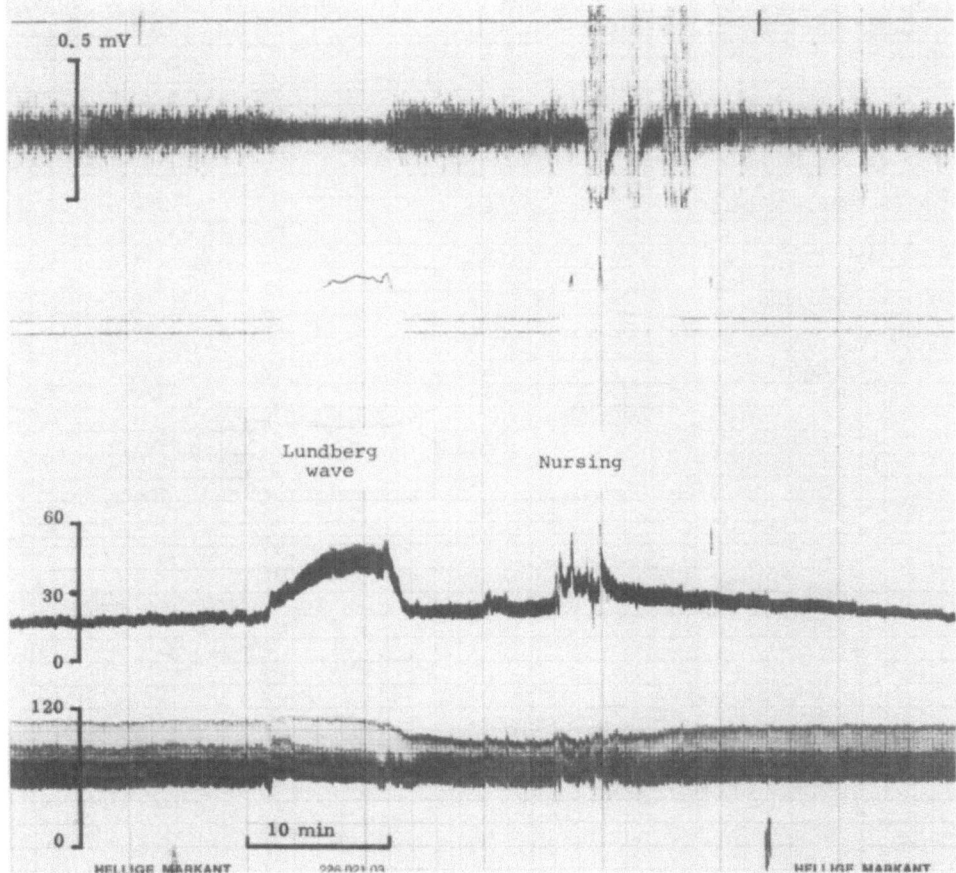

Fig. 1   Lundberg wave and pronounced ICP rise due to nursing procedures in an 11 year old boy with temporobasal contusion following a sledge accident. Traces from top to bottom: one-channel EEG; intracranial pressure, arterial pressure

In three adult patients thiopentone infusion was initiated at a rate of 80 mg/min for half an hour, followed by 8 mg/min as constant rate infusion, which, according to pharmacokinetic calculations (13), should yield and maintain burst-suppression from the start of thiopentone medication.

Burst-suppression EEG was considered the lower range of barbiturate coma even in ICPs irresponsive to the therapy. Thiopentone infusion was interrupted when ICP could be kept in the normal range for 24 hours or when cerebral perfusion pressure remained zero for 24 hours.

Table 1   Clinical features of 27 patients treated with thiopentone infusion for otherwise intractable raised ICP

	Survivors: 13		Non-survivors: 14	
Age	16,5 $\pm$ 6,5		24,0 $\pm$ 10,1	
Coma score	5,6 (3-10)		5,0 (3-6)	
Intracerebral haematoma	5	38%	6	43%
Contusion	9	69%	9	64%
Diffuse swelling	4	31%	2	14%
Skull fractures				
depressed	2	15%	1	7%
other	6	46%	8	57%
Pupillary anomalies	4	31%	7	50%
bilat. fixed	-	-	2	14%
Convulsions	1	8%	1	7%

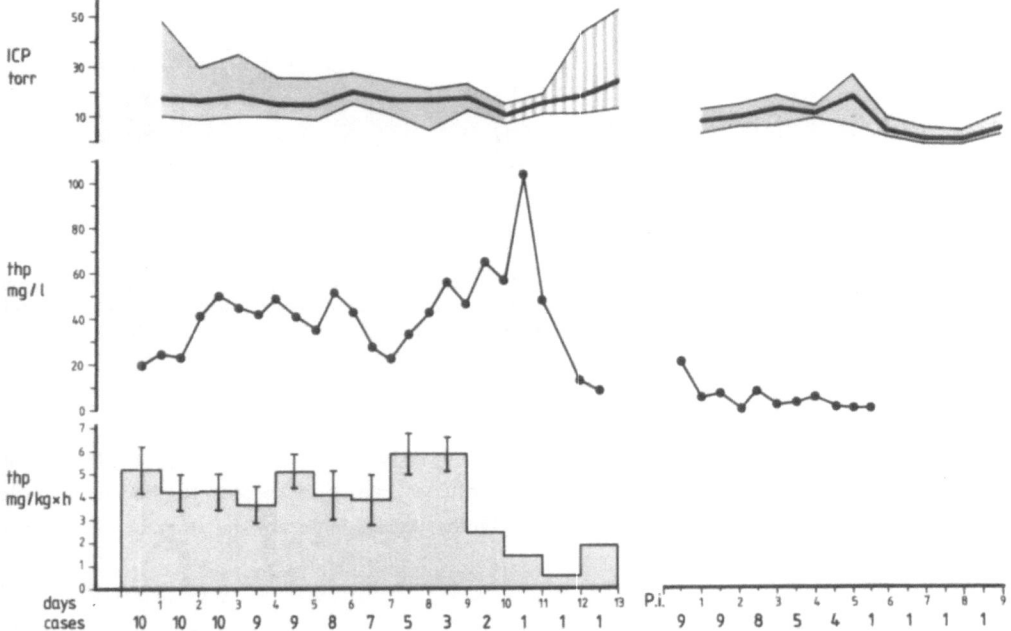

Fig. 2   Intracranial pressures, serum thiopentone levels and thiopentone infusion rates (median) in 10 survivors during (first part) and after (second part) therapy. Straight line: mean daily ICP, shaded area: daily range of ICP. Thiopentone levels estimated at 6 a.m. and 6 p.m.

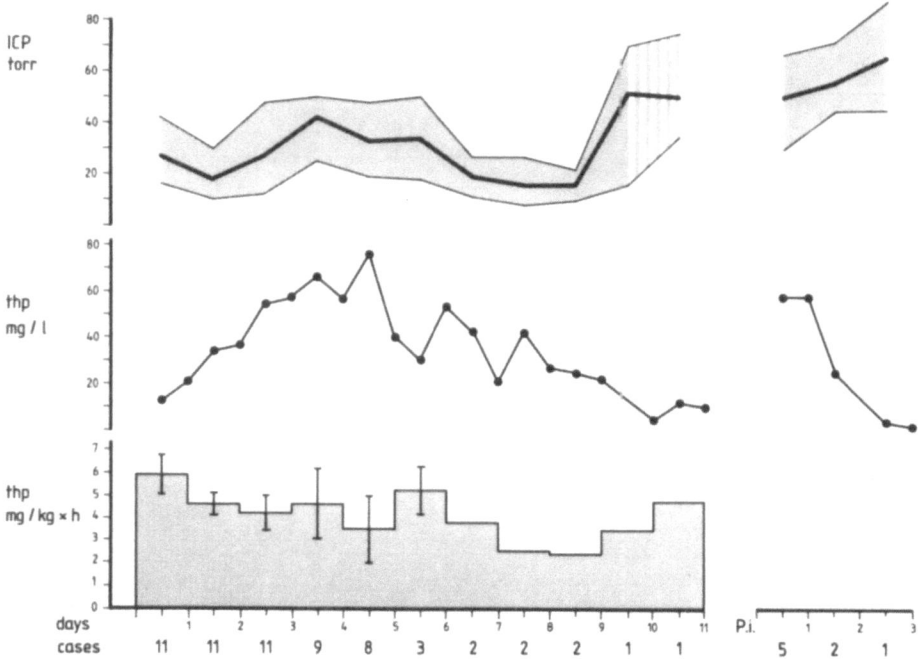

Fig. 3   ICP serum thiopentone levels and thiopentone infusion rates in 11 non-survivors. Symbols as in Fig. 2

Results

The clinical course regarding ICP, thiopentone plasma levels and thiopentone infusion rates in 21 patients, in whom these data were collected consistently, is shown in Fig. 2 and Fig. 3. ICP in survivors was controlled fairly well during infusion, but in some patients during the second week of treatment dosage had to be increased due to recurrent elevations of intracranial pressure.

In three individuals, thiopentone coma had to be reinstituted after two days for a second course resulting in thiopentone serum levels of 63 and 103 mg/l in the two patients treated the longest.

When infusion was discontinued, serum thiopentone levels fell to zero within 4 days, mean intracranial pressure rose slightly, due to loss of sedation and enhanced pressure response to nursing procedures.

In the non-survivors (Fig. 3), intracranial pressure was reduced only slightly by thiopentone, but in most patients eventually could not be controlled despite increasing infusion rate and

Table 2 Observations during thiopentone treatment of 27 patients with raised ICP. (Wilcoxon tests for paired and unpaired values for differences in ICP's in survivors and non-survivors, before and after treatment; ←——→differences significant at $a \leqslant 0.05$)

	13 survivors	14 non-survivors
thp. infusion time (days)	$7{,}3 \pm 2{,}5$	$5{,}7 \pm 1{,}5$
ICP start thp. (torr)	$35 \pm 10{,}1$	$44 \pm 19{,}4$
ICP end thp. (torr)	$14 \pm 9{,}32$	$64 \pm 30{,}4$
EEG burst-suppression:	n = 10	n = 11
Hours after thp. start:	10,25 (5-85)	28 (0,25-94)
Serum thp. (mg/l)	24,3 (12,8-51,6)	27,9 (20,8-62,1)
ICP (torr)	$27{,}5 \pm 11{,}8$	20 ± 14

rising plasma barbiturate levels, which in one individual reached 183 mg/l. In fact, this graph presents two different groups regarding the reaction of ICP to barbiturate infusion. In five patients, ICP rose during three to four days to equal mean arterial pressure; in four patients, however, after successful control of ICP for up to five days a secondary rise occurred relatively suddenly despite continued barbiturate dosage thus abolishing cerebral perfusion pressure. In the end, eight patients died of global cerebral ischemia. We, however, never diagnosed isolated brain death prior to barbiturate levels having fallen to zero.

Further observations are summarized in Table 2.

Effective control of intracranial pressure can be demonstrated in survivors. Medians of intracranial pressures at the start of treatment were not significantly different in the two groups. However, their fall in survivors and their rise in non-survivors towards the end of treatment are significant as well as the difference between the two groups at that time. Burst-suppression patterns in the EEG occurred at very different times in the two groups depending on the thiopentone dosage scheme, and at mean plasma levels slightly lower than those normally necessary to produce loss of corneal reflex (3). These, however, had to be increased to values from 45 to 74 mg/l to maintain this EEG pattern during treatment, which resembles acute tolerance to thiopentone (1,21).

Appearance of burst suppression was not related to a particular range of intracranial pressure, nor did maintenance of this pattern prevent further pressure increases in non-responders.

Table 3   Outcome (11) in 27 brain trauma patients treated by thiopentone
           infusion for raised ICP

Grade	Term	Own patients
1	Good recovery	6
2	Moderately disabled	2
3	Severely disabled	3
4	Vegetative	2[a]
5	Dead	14

[a] 4 weeks after treatment

The outcome was evaluated by the Glasgow outcome scale (Table 3) in survivors at least six months after treatment. However, in two patients the vegetative state does not resemble final outcome, since barbiturate therapy had been withdrawn only four weeks previously.

Barbiturate coma in brain trauma patients may entail severe complications, but attention has mainly focused on possible cardiovascular impairment.

In our patients, reductions in systemic blood pressure were seen during rapid bolus injection or in attempts to achieve burst suppression too urgently, but were easily corrected by blood volume expansion and dobutamine infusion 0.2 mg/kg . h in every instant, even during administration of 80 mg/min for half an hour (Fig. 4).

In one survivor, however, severe cholostatic jaundice occurred with hyperthermia and exanthematous rashes, and serum bilirubin of 16.5 mg/100 ml on the fourth day following 13 days of barbiturate treatment.

Excessive extra- and intrahepatic cholostasis was confirmed by laparotomy as well as by microscopic sections. This course is similar to thiopentone hepatitis (10) due to allergic reactions to this compound (18).

Two survivors developed ulcerative reflux oesophagitis during the first week of barbiturate coma, which in one case necessitated creation of a gastrostoma for successful treatment. De los Reyes (5) reported silent duodenal ulceration on the 16th day of pentobarbitone coma in cerebral trauma. Whether this complication is peculiar to barbiturate treatment or whether the problem is obtundation of symptoms of gastrointestinal ulceration known frequently to occur in brain trauma, remains to be decided.

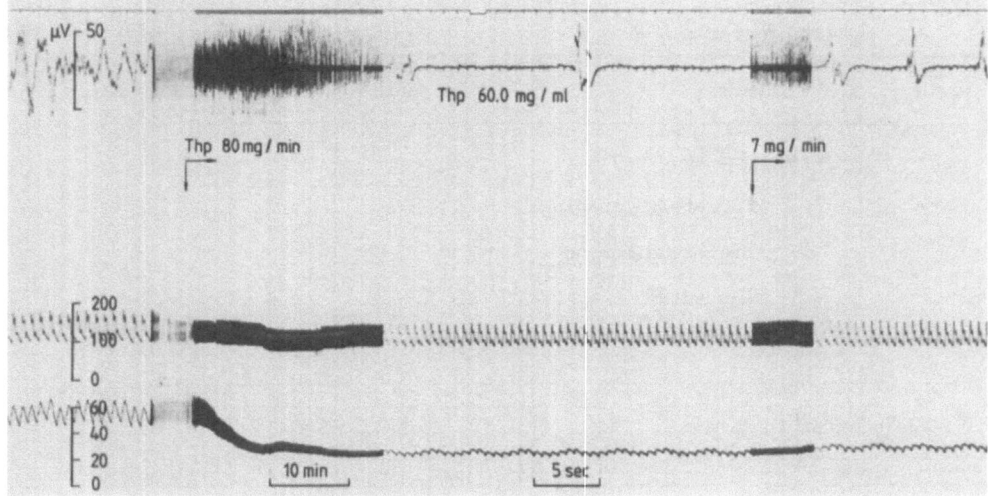

Fig. 4   27 year old patient (75 kg) with left parietal contusion and intracerebral clot. Traces from top to bottom: EEG, ICP, arterial blood pressure. Note effectively lowered ICP without major changes in systemic pressure at thiopentone 80 mg/min for 30 minutes, followed by 8 mg/h; dobutamine 0.2 mg/kg . h was infused concomitantly

Discussion

The benefit of thiopentone infusion to patients with raised intracranial pressure is not obvious in our series. Relating coma score to outcome after brain trauma, Jennett et al. (12) showed that 87% of grade 3/4 patients and 53% of grade 5/6/7 patients not treated with barbiturates die or remain vegetative, whereas in our patients this occurred in 66% and 69% respectively.

Bruce et al. (4) and Marshall et al. (14) have reported far more favourable outcomes in their patients with barbiturate coma (Table 4).

However, in the Philadelphia series mass lesions were only 23% in overall patients as compared to 40% in the present communication, and diffuse swelling occurred in 34% in contrast to 22%. The mean age of those patients was lower, although we must concede that our survivors are significantly younger than our non-survivors. Our patients were more severely impaired and were older than in the Philadelphia series, both factors adding to unfavourable outcome in brain trauma.

The difference from Marshall's series is not readily explained, wherein only one of six patients with surgically treated intracranial mass died, despite their worse neurological state, as evidenced by 44% of cases exhibiting bilaterally unreactive pupils and 48% of cases presenting as decerebrate.

Moreover, the mean age of the San Diego patients was distinctly higher.

In our patients on CT scan, intracerebral hematomas were not interpreted as amenable to surgical evacuation, but although only three of the six non-surviving hematomas died with intractable intracranial pressure, evacuation might have to be attempted in some.

On the other hand, among the deaths not attributable to raised intracranial pressure there were two cardiac failures and one death due to meningitis. Three patients died with hyperthermia, which was of central origin in two patients and due to sepsis in one patient.

An even more distressing result with 65% mortality in 20 cases despite barbiturate coma, was reported by Frost et al. in 1981. However, their series comprised subarachnoid hemorrhage and gunshot wounds as well as REYE's syndrome. In their trauma cases, mortality even amounted to 82%.

Since Gennarelli et al. (7) demonstrated that outcome from severe head injury is markedly influenced by the type of lesion, not only our series is definitely too small to allow categorization into the seven lesion types necessary to decide on this issue, and thus not only in our own patients may it prove difficult to demonstrate definite benefit of

Table 4   Comparison among the Philadelphia (4), San Diego (_4) and Heidelberg series of outcome in brain trauma patients with raised ICP treated by barbiturate infusion

	Philadelphia 1979	San Diego 1979	Heidelberg 1982
Cases	23	25	27
Age	2-17	32	21 (3-45)
Results:			
Group I	65%	48%	30%
Group II	9%	16%	18%
Dead	26%	36%	52%
Dead with ICP = MABP	50%	55%	57%

Group I: good recovery, moderately impaired
Group II: severely impaired, vegetative

barbiturate coma in brain trauma with raised intracranial pressure. In the single patient, a reduction of initially high intracranial pressure may ultimately be followed by a favourable outcome, but in a huge percentage of non-survivors intracranial pressure may be lowered effectively at first, only to increase inexorably to levels compromising cerebral perfusion.

Undoubtedly, thiopentone and other barbiturates will reduce intracranial blood volume by diminishing cerebral blood flow secondary to lowered oxygen demand with depressed cerebral functional metabolism (15). Thus, the drug may be used most promisingly to lower raised intracranial pressure in cases of cerebral hyperperfusion after brain trauma.

However, evidence is difficult to obtain in human brain trauma as to whether there are "cerebroprotective" effects of barbiturates other than those produced by lowering raised intracranial pressure. Moreover, more recent experimental results have cast doubt on barbiturate effects previously claimed for protection from cerebral ischemia and/or anoxia.

Membrane stabilization has been discredited as a barbiturate effect by the work of Astrup et al. (2), who in the dog showed that postischemic potassium efflux from the brain is not to be prevented by thiopental. Free radical scavenging by thiopental, demonstrated in vitro, does not seem to be a mechanism for protection from focal or global cerebral ischemia in vivo (19).

In the monkey neck tourniquet model of global cerebral ischemia, first promising results indicating influences by barbiturates on postischemic neurologic deficit scores could not be reproduced in the same laboratory (8).

In a cat model of global cerebral ischemia brought about by ventricular fibrillation (20), there were no differences in neurologic deficits among thiopentone-treated and untreated survivors.

In the same study, however, the significantly fewer deaths in the thiopentone group were correlated with a significantly lower incidence of seizure patterns in the EEG.

Thus Overgaard's suggestion for early use of barbiturate sedation in brain trauma seems to gain additional support (17). He demonstrated marked reduction in frontoparietal regional cerebral blood flow in brain trauma patients and cited Nilson's (16) conclusion from work in rats that brain trauma may be termed an excitatory event. He thus pointed to the probable disparity of cerebral metabolic demands and supply of oxygen and substrates, which might be ameliorated by metabolic depressants.

However, with the probable exception of reducing seizure-like activity in brain trauma and its underlying cerebral hypermetabolism and concomitant hyperperfusion (Fig. 5), we believe

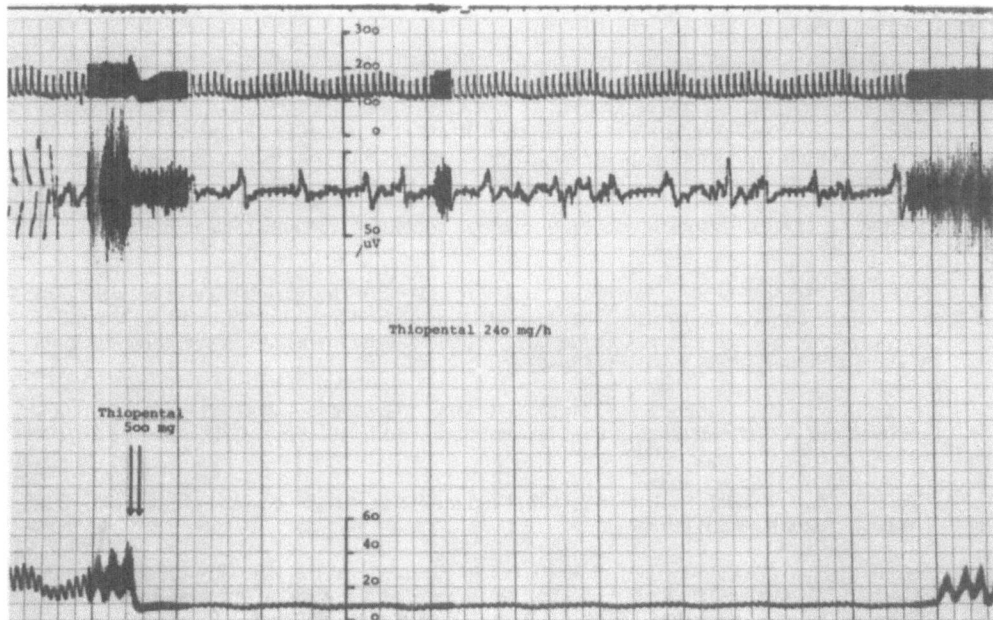

Fig. 5   42 year old patient (70 kg) with right frontal cerebral contusion and two cardiac arrests successfully treated during transport; observation 6 h after arrival. Traces from top to bottom: arterial pressure, EEG, ICP. Rhythmic seizure activity (left part), concomitant ICP elevations due to synchronously increased cerebral blood flow successfully suppressed by 500 mg thiopentone, constant rate infusion of 240 mg/h being uneffective

that in brain trauma patients induction of barbiturate coma should be restricted to raised intracranial pressure because of

1.   severe side effects to be expected in some cases
2.   as yet ambiguous evidence of real amelioration in outcome
3.   insufficient evidence of cerebral protective action.

In presenting these results, we are aware that newer hypnotic substances will possibly exhibit fewer side effects, especially as regards systemic circulation, and that this is another anecdotal report. But controlled trials, however necessary, have not yet overcome organization obstacles, and even emotional restraints.

References

1. Altenburg BM, Michenfelder JD, Theye RA (1969) Acute tolerance to thiopental in canine cerebral oxygen consumption studies. Anesthesiology 31:443-448

2. Astrup J, Skovsted P, Gjerris F, Sørensen HR (1981) Increase in extracellular potassium in the brain during circulatory arrest: effects of hypothermia, lidocaine and thiopental. Anesthesiology 55:256-262

3. Becker KE (1978) Plasma levels of thiopental necessary for anesthesia. Anesthesiology 49:192-196

4. Bruce DA, Raphaely RA, Swerdlow D, Schut L (1980) The effectiveness of iatrogenic barbiturate coma in controlling increased ICP in 61 children. In: Shulman K, Marmarou A, Miller JD, Becker DP, Hochwald GM, Brock M. Eds. Intracranial pressure. Springer, Berlin Heidelberg New York, pp 630-632

5. De los Reyes RA, Babcock RA, Malik GM, Diaz FG, Ausman JI (1981) Silent duodenal perforation: A difficult diagnosis in iatrogenic barbiturate coma. Crit Care Med 9:104-105

6. Frost EAM, Tabaddor K, Kim BY (1981) Induced barbiturate coma: results in 20 cases (Abstr.). Anesth Analg 60:247

7. Gennarelli TA, Spielman GM, Langfitt TW, Gildenberg PL, Harrington T, Jane JA, Marshall LF, Miller JD, Pitts LH (1982) Influence of the type of intracranial lesion on outcome from severe head injury. A multicenter study using a new classification system. J Neurosurg 56:26-32

8. Gisvold SE, Safar P, Hendricks H, Alexander H (1981) Thiopental treatment after global brain ischemia in monkeys. Anesthesiology 55:A97

9. Gobiet W (1977) Intensivtherapie nach Schädel-Hirn-Trauma. Springer, Berlin Heidelberg, New York, p 48

10. Hasselström L, Kristoffersen MB (1979) Hepatitis following thiopentone. A case report. Br J Anaesth 51:801-804

11. Jennett B, Bond MR (1975) Assessment of outcome after severe brain damage. Lancet 1:480-484

12. Jennett B, Teasdale G, Braakman R, Minderhoud H, Heiden J, Kurze T (1979) Prognosis of patients with severe head injury. Neurosurgery 4:283-289

13. Lauven PM Personal communication. Department of Anesthesiology, University of Bonn, FRG

14. Marshall LF, Smith RW, Shapiro HM (1979) The outcome with aggressive treatment in severe head injuries. Part II Acute and chronic barbiturate administration in the management of head injury. J Neurosurg 50:26-30

15. Michenfelder JD (1974) The interdependency of cerebral functional and metabolic effects following massive doses of thiopental in the dog. Anesthesiology 41:231-236

16. Nilsson B, Nordström CH (1977) Rate of cerebral energy consumption in concussive head injury in the rat. J Neurosurg 47:274-281

17. Overgaard J, Mosdal C, Tweed WA (1981) Cerebral circulation after head injury. Part 3: Does reduced regional cerebral blood flow determine recovery of brain function after blunt head injury? J Neurosurg 55:63-74

18. Pagliaro L, Campesi G, Aguglia F (1969) Barbiturate jaundice. Report of a case due to a barbital-containing drug with positive rechallenge to phenobarbital. Gastroenterology 56:938-943

19. Steen PA, Michenfelder JD (1980) Mechanisms of barbiturate protection. Anesthesiology 53:183-185

20. Todd MM, Chadwick HS, Shapiro HM, Dunlop BJ, Marshall LF, Dueck R (1982) The neurologic effects of thiopental therapy following experimental cardiac arrest in cats. Anesthesiology 57:76-86

21. Toner W, Howard PJ, McGowan WAW, Dundee JW (1980) Another look at acute tolerance to thiopentone. Br J Anaesth 52:1005-1008

Brain Resuscitation in Children: Current Indications and Future Directions

D. A. Bruce

Children's Hospital of Philadelphia, University of Pennsylvania,
School of Medicine, Philadelphia, PA, USA

The concept of brain resuscitation arose from the observations that cardiopulmonary resuscitation was frequently effective in restarting the arrested heart, but that severe residual brain damage frequently interfered with complete recovery of the patient. Brain resuscitation is, however, in its broadest sense the application of intensive care measures to patients with brain injuries, for whatever cause, to maximize cerebral recovery. In children, the most frequent pathological conditions requiring such measures are head trauma, anoxia or ischemia, hemorrhagic encephalitis, and Reye's syndrome. Acute stroke and intracranial hemorrhage are less common indications in children.

It is clear from the name, brain resuscitation, that the concept is not to prevent primary injury to the brain, since this has already occurred prior to any medical involvement, but to prevent new injury and maximize recovery. There are three major mechanisms by which brain resuscitation may be helpful.

A.   Control of intracranial pressure (ICP)

B.   Matching CBF to $CMRO_2$

C.   Preventing biochemical interactions that lead to secondary cell damage

Control of the ICP will prevent cerebral herniation syndromes and will prevent secondary episodes of cerebral ischemia locally or globally due to reduced cerebral perfusion pressure. It is becoming increasingly obvious that increase in metabolic demand (e.g. seizures in patients with other brain injuries) can lead to destruction of neural function and secondary injury. Thus, measures designed to match the cerebral metabolism to available cerebral blood flow, or vice versa, may be effective in preventing secondary damage from mismatch between metabolic demand and substrate delivery. Recent experimental data strongly suggest that control of abnormal electrical activity in post-ischemic brain is beneficial and improves survival (9). Finally, there is the question whether there are biochemical reactions set in motion by the primary pathological event, ischemia or trauma, that are progressive or, indeed, begin post-lesional cerebral resuscitation. If so, then it might be possible to prevent a secondary biochemical lesion in the brain. A variety of biochemical reactions have been targeted but none actually proven to produce secondary damage. Recent data does infer that some process occurs that produces secondary damage in the post-ischemic reperfusion period (8). Despite this, the only agents shown to be "protective" against progressive hypoxia or ischemia have been agents that decrease cerebral metabolism and flow, e.g. hypothermia, barbiturates (5) and isoflurane (6). Studies demonstrating beneficial effects of naloxone in stroke models have been published (8). Calcium blockers have not yet been proven to be beneficial despite hopes that they would prove effective agents (4). In reviewing our clinical experience in children, we shall examine the evidence for the mechanisms discussed above and in which circumstances cerebral resuscitation has proved to be beneficial.

Head injuries

The second head injury is now a well-accepted phenomenon produced either by hypoxia, ischemia or increased ICP. The goal of current management of head-injured patients is to prevent this secondary injury by the early establishment of a good airway, careful fluid management, and control of blood pressure and ICP within the normal range. Early mortality in children is associated with severe intracranial hypertension and all of the children in our series who have died primarily had an initial ICP of 40 torr or greater despite intubation and hyperventilation. Delayed damage due to secondary ischemia still ocurs unless aggressive therapy to control ICP is begun immediately. The concepts A and B of cerebral resuscitation described above are valuable in the treatment of children with head injuries. The use of muscle paralysis, controlled ventilation, mannitol, barbiturates and hypothermia to maintain a normal ICP and to match cerebral blood flow and metabolism have significantly improved the outcome (1,2). As yet, there is no evidence of a chemical "protective" effect of barbiturates or any other drug in head injury. With current best management, children with a Glasgow Coma Score above 5 should not die from their head injury. Children who are decerebrate or decorticate, Glasgow Coma Score 4 to 5, have a current mortality of 13% and those who are flaccid, of 50%. Equally important, the percentage of good recovery and moderate disability should be anywhere from 40-80% depending on the initial Glasgow Coma Score. This is a situation where definite secondary injury can be prevented. The results of intensive therapy are worst in those children who show extremes of primary axonal damage on CT scan (10). In these children, mortality and morbidity are the highest, probably because most of the injury is primary and, therefore, despite the prevention of secondary injury by methods of resuscitation an alteration of outcome cannot be achieved.

Near drowning

We have compared the outcome of near drowning victims in the Children's Hospital of Philadelphia over two separate time periods, 1976 - 1979 and 1980 - 1982. During the first period, all patients in coma received endotracheal intubation, controlled ventilation and ICP monitoring. The ICP was rigidly controlled below 20 torr, if possible, with therapies such as

Table 1   Outcome of near drowning in childhood

	ICP monitored & treated with barbs to ICP↓ 20	ICP not monitored no specific therapy
Total	17	17
Absent calorics	8	10
Fixed dilated pupils	11	13
Died	8	8
G.R./M.D.	2	6
S.D./Veg.	7	3

steroids, mannitol, lasix, and, if necessary, barbiturates (Nembutal) in coma doses. In the second time period, the same management of airway and ventilation was continued, the ICP was not monitored, no specific therapy was given for ICP control and the only barbiturate used was phenobarbital as an anti-seizure medication in three patients. As shown in Table 1, the mortality was the same but the number of patients who recovered well was greater in the second time period. The reason for this change in management was that all children in the first time period who developed intracranial hypertension either died, were severely disabled or vegetative. While it was possible to control the ICP in all but two patients, there was no evidence that ICP control improved outcome. Indeed, we conclude that the findings of increased ICP were secondary to neocortical neuronal swelling and were a reflection of an irreversible insult. This seems to be borne out by the results in the last two years which show at least a good or better recovery rate when no specific therapy directed towards ICP control or manipulation of metabolism or brain protection is used. While we have not used barbiturates primarily in this group of patients as protective agents, our results are no worse than those situations in which barbiturates have been used in warm water drowning (7). The situation suggests that anoxia produces significant primary injury and that it is the degree of this primary injury that decides the outcome. Thus, brain protection after the event with the current state of the art appears to be incapable of altering outcome.

Reye's syndrome

Reye's syndrome is a toxic encephalopathy associated with acute fatty infiltration of the liver. The mortality in the past has been as high as 80%. The introduction of good ICU care and control of ICP has lowered the mortality for children in coma to 20-50%. Many of the deaths occurred with evidence of elevated ICP and, over the last few years, therapeutic efforts have been directed to prevent elevated ICP and, hopefully, to protect the brain in

Table 2 Outcome in children with reyes syndrome

	Group A (n = 20)	Group B (n = 23)
Good recovery	6 (30%)	11 (48%)
Impaired	4 (20%)	4 (17%)
Dead	10 (50%)	8 (35%)

Group A

Nembutal only in response to ICP over 20

Group B

Immediate therapeutic Nembutal coma

children with severe Reye's syndrome. At the Children's Hospital of Philadelphia, 43 children with severe Reye's syndrome were treated and ICP rigidly controlled. The children were divided into two groups, A and B. In group A, ten of the 20 children received therapeutic Nembutal coma because ICP could not be controlled below 20 torr by other means. In group B, all 23 patients received immediate high-dose barbiturate coma and hypothermia to "protect" the brain and control ICP. Table 2 (3) shows the outcome in these two groups. While the trend is towards better outcome, there was no statistical difference between the groups. There was, however, a higher incidence of pneumonia (43% versus 10%) in group B. Recent studies of CBF and metabolism in our intensive care unit have shown that the outcome of Reye's syndrome may be predicted on the basis of dissociation between cerebral blood flow and cerebral metabolic rate. Children with this pattern appear to all die. This suggests that the elevated ICP may not be the cause of death, but the result of extensive neuronal death and that the deciding factor in outcome is the severity of the encephalopathy. This may explain the failure of early barbiturates to have any beneficial effect. Again, this suggests that in a disease state in which the majority of the injury is primary, the role of brain resuscitation is minor.

Discussion

Current concepts in the application of methods of brain resuscitation in children have been beneficial in altering the outcome from severe head injury. These results are due to the prevention of significant secondary brain injury due to ischemia or hypoxia. In Reye's syndrome, while aggressive pursuit of ICP control has improved outcome, there is no evidence that any chemical protective therapy further improves outcome and we suggest that the poor outcome is a function of the severity of the encephalopathy and the degree of primary damage produced. In near drowning, the outcome does not appear to be improved by

162

methods of brain resuscitation. Clearly, better initial CPR and improved pulmonary care will contribute to a large number of children reaching hospital alive, but the deciding factor still appears to be the amount of primary brain injury. Our conclusion is that there is no clinical evidence, in children, that brain protection is currently available as a therapeutic modality. Aspects A and B of brain resuscitation, ICP control and match of metabolism to blood flow may, indeed, alter outcome in those states in which cerebral herniation or alterations of cerebral blood flow due to altered perfusion pressure result in secondary damage.

The future of brain resuscitation will depend upon the identification at a basic science level of mechanisms that are responsible for secondary biochemical damage. There can then be developed therapeutic regimens to alter or prevent the damage and these regimens may then be introduced in a controlled fashion to clinical practice. The initial introduction should be in the way of randomized controlled studies. The best management currently available varies depending upon the disease process, but there is no clinical evidence that chemical brain protection is a viable or beneficial goal of therapy at the present time. Re-establishment of adequate cerebral blood flow, delivery of substrate, control of seizures and control of the general body physiology would appear the best that can be done.

References

1. Bruce DA, Alavi A, Bilaniuk LT, Dolinskas C, Obrist WA, Zimmerman RA, Uzzel B (1981) Diffuse cerebral swelling following head injuries in children: The syndrome of "malignant brain edema". J Neurosurg 54:170-178

2.. Bruce DA, Schut L, Bruno LA, Wood JH, Sutton LN (1978) Outcome following severe head injuries in children. J Neurosurg 48:679-688

3. Frewin TC, Swedlow DB, Watcha M, Raphaely RC, Godines RI, Heiser MS, Kettrick RG, Bruce DA (1982) Outcome in severe Reye's syndrome with early pentobarbital coma and hypothermia. J Peds 100:663-665

4. Gergis SD, Sokoll MD, Sarracino SM, Kassell NF (1982) Effect of a Ca^{++} channel blocker on cerebral reflow phenomenon in the dog. Anesthesiology 57:A363

5. Michenfelder JD, Theye RA (1973) Cerebral protection by thiopental during hypoxia. Anesthesiology 39:510

6. Neuberg LA, Michenfelder JD (1982) Cerebral protection by isoflurane during hypoxemia or ischemia. Anesthesiology 57:A335

7. Oakes DO, Sherck JP, Maloney JR, Charters AC (1982) Prognosis and management of victims of near-drowning. J Trauma 22:544-549

8. (1982) Protection of the brain from hypoxia. J Cereb Blood Flow Metabol 2: Suppl 1:1-101

9. Todd MM, Chadwick HS, Shapiro HH, Dunlop BJ, Marshall LF, Dueck P (1982) The neurological effects of thiopental therapy following experimental cardiac arrest in cats. Anesthesiology 57:78-86

10. Zimmerman RA, Bilaniuk LT, Gennarelli T (1978) Computed tomography of shearing injuries of the cerebral white matter. Radiology 127:393-396

Statement and Conclusions

In an attempt to draw conclusions from the material presented at this meeting, the following views in a final discussion were put forward. With the exception of therapeutic procedures, they were intended to reflect the opinions shared by the participants.

Irreversible membrane depolarization and its deleterious effects on cellular metabolism are believed to be primary factors in the development of irreversible brain death. Thus, structurally visible neuronal damage seems to require not only energy failure to a degree that severely affects ion and water compartmentation, but also irreversible membrane disruption. Protection from irreversible brain death would mean prevention of membrane failure or irreversible membrane depolarization and improvement of cellular metabolism, particularly energy production. Lactic acidosis probably plays a role in accelerating or potentiating membrane damage in addition to exaggerating ionic and water imbalances. The events initiating and propagating membrane dysfunction have yet to be positively identified, but may well involve intracellular movement of calcium ions.

The role of cerebral perfusion in the process of ultimate neuronal death has yet to be elucidated. The selective vulnerability of the brain is evident in ischaemic penumbra, which is well defined in acute ischaemia. Whether this penumbra can persist in chronic regional ischaemia, or whether there is a definite border between viable and necrotic tissue needs clarification.

Failure in synaptic transmission mechanisms as opposed to failure of cerebral oxidative metabolism may be partly responsible for neurologic dysfunction after cerebral ischaemia. The ten-fold increase in rat cerebral cortex cyclic-AMP accumulation in response to noradrenaline during ischaemia and its delayed attentuation after resuscitation may correlate with the early hyperactive phase followed by a delayed deterioration of cerebral function in postischaemic encephalopathy.

Apparent discrepancies in the studies on the cerebral protective effects of barbiturates after global ischaemia are likely attributable to differences in methodology, especially the degree of hypermetabolic stress in controls which can be markedly affected by anaesthetic agents and other drugs. Free fatty acid liberation during cerebral ischaemia appears to correlate with the development of ischaemic injury. The relative magnitude of FFA attentuation by phenytoin and barbiturates indicates that these drugs are of greatest potential effectiveness in ameliorating ischaemic brain damage. Furthermore, for maximal therapeutic effects, it appears to be unnecessary to exceed anaesthetic or anticonvulsant doses or to obtain an isoelectric EEG. These findings also suggest that the efficacy of cerebral drugs is not primarily related to their anaesthetic or cerebral metabolic depressant effects. These latter views, however, were not shared by all participants. Some felt effectiveness of barbiturates in global ischaemia was not proven save in postischaemic convulsive activities. With this latter mechanism accepted, isoelectric EEG and maximum suppression of oxidative metabolism indeed would seem necessary.

The time of application of cerebral protective measures after cerebral ischaemia also appears to be crucial. Experimentally, immediate barbiturate treatment only increased survival. Similar effects of moderate hypothermia (28°C) were lost when normothermia was restored.

Hypnotics, considered as cerebral protective drugs after focal or global ischaemia, may exert variable effects on cerebral metabolism. Thus, using the magnitude of the increase in rat cortical gray matter lactate as an indicator of brain dysfunction postischaemia, gamma-hydroxy-butyrate proved to be more effective than thiobarbiturate.

A calcium antagonist in a dog model of ventricular fibrillation followed by 6 hours of recirculation had no appreciable effects on the extent of neurologic damage. Cerebrospinal fluid levels of inorganic phosphate and creatine phosphokinase correlated poorly with severity of brain damage.

The difficulties encountered in demonstrating the effectiveness of pharmacotherapy in experimental models of cerebral ischaemia indicate that it may be even more difficult to demonstrate in clinical trials. Justification for clinical trials must be based on improved quality of outcome, not only improved survival. Categorization of the patterns and severity of ischaemia and traumatic brain damage is a prerequisite for comparability of outcome and hence, of effective treatment. Patient age must also be considered as one of the important factors influencing outcome. Treatment must be standardized, but this will limit the feasibility of clinical trials even more than the stratification of patients. In stroke patients, both problems are farthest from being solved due to variable damage categories and different treatment modalities. Reversibility of stroke-induced cerebral damage requires careful evaluation before entering a patient into the study.

Presently, raised intracranial pressure is the only definite indication for treatment with hypnotics such as barbiturates, etomidate, althesin or gamma-hydroxybutyrate in brain trauma patients. Most participants consider that dosage should be restricted to effective control of intracranial pressure to a normal range of 15 torr rather than EEG patterns. Clinical experience has shown that with high serum levels of some, if not all of these substances, cardiovascular, gastrointestinal and infectious complications are likely to increase.

Accepting the ICP as a guide to pharmacotherapy by hypnotics for brain trauma, it must be kept in mind that although ICP is easily measured, like systemic blood pressure, it yields only limited information on haemodynamics and metabolic changes. Thus, more attention should be paid to arteriovenous differences for oxygen and metabolites to gain more insight into the effects of cerebral protective therapy. The goal of this therapy is to preserve and restore normal cerebral function.

Subject Index

Adrenal Actions on Brain

Editors: D. Ganten, D. Pfaff
1982. 25 figures. V, 153 pages
(Current Topics in Neuroendocrinology, Volume 2)
ISBN 3-540-11126-3

Brain Abscess and Meningitis Subarachnoid Hemorrhage: Timing Problems

Editors: W. Schiefer, M. Klinger, M. Brock
1981. 219 figures, 134 tables. XIX, 519 pages
(Advances in Neurosurgery, Volume 9)
ISBN 3-540-10539-5
Distribution rights for Japan: Nankodo Co. Ltd., Tokyo

Cerebral Aneurysms

Advances in Diagnosis and Therapy
Editors: H. W. Pia, C. Langmaid, J. Zierski
1979. 265 figures, 143 tables. XIV, 458 pages
ISBN 3-540-09159-9

Cerebrovascular Transport Mechanisms

International Congress of Neuropathology,
Vienna, September 5–10, 1982
Editors: K.-A. Hossmann, I. Klatzo
1983. 41 figures. 160 pages
(Acta Neuropathologica, Supplement 8)
ISBN 3-540-12204-4

Microsurgery for Cerebral Ischemia

Editors: S. J. Peerless, C. W. McCormick
1980. 282 figures. XVII, 372 pages
ISBN 3-540-90495-6

Microvascular Anastomoses for Cerebral Ischemia

Editors: J. M. Fein, O. H. Reichman
1978. 197 figures, including 12 four-color plates. XV, 324 pages
ISBN 3-540-90240-6

Neuroradiology

A Neuropathological Approach
By R. Kautzky, K. J. Zülch, S. Wende, A. Tänzer
Translated from the German edition by W. M. Boehm
1982. 251 figures. XII, 324 pages
ISBN 3-540-10934-X

Spontaneous Intracerebral Haematomas

Advances in Diagnosis and Therapy
Editors: H. W. Pia, C. Langmaid, J. Zierski
1980. 292 figures. XV, 415 pages
ISBN 3-540-10146-2

Treatment of Cerebral Edema

Editors: A. H. Hartmann, M. Brock
1982. 95 figures, 30 tables. X, 176 pages
ISBN 3-540-11751-2

Springer-Verlag
Berlin
Heidelberg
New York
Tokyo

Experimental Brain Research

Supplement 1:
Afferent and Intrinsic Organization of Laminated Structures in the Brain
(7th International Neurobiology Meeting)
Editor: O. Creutzfeldt
1976. 127 figures. XXIII, 579 pages
ISBN 3-540-07923-8

Supplement 2:
Hearing Mechanisms and Speech
EBBS-Workshop, Göttingen, April 26–28, 1979
Editors: O. Creutzfeldt, H. Scheich, C. Schreiner
1979. 85 figures, 12 tables. XXIII, 413 pages
ISBN 3-540-09655-8

Supplement 3:
Gonadal Steroids and Brain Function
IUPS-Satellite-Symposium, Berlin, July 10–11, 1980
Editors: W. Wuttke, R. Horowski
1981. 136 figures, 10 tables. XIII, 373 pages
ISBN 3-540-10606-5

Supplement 4:
The Renin Angiotensin System in the Brain
A Model for the Synthesis of Peptides in the Brain
Editors: D. Ganten, M. Printz, M. I. Phillips, B. A. Schölkens
1982. 108 figures, 46 tables. XVII, 385 pages
ISBN 3-540-11344-4

Supplement 5:
The Aging Brain
Physiological and Pathophysiological Aspects
Editor: S. Hoyer
1982. 52 figures, 66 tables. XIV, 281 pages
ISBN 3-540-11394-0

Supplement 6:
The Cerebellum – New Vistas
Editors: S. L. Palay, V. Chan-Palay
1982. 264 figures, 16 tables. XVII, 637 pages
ISBN 3-540-11472-6

Supplement 7:
Neural Coding of Motor Performance
Editors: J. Massion, J. Paillard, W. Schultz, M. Wiesendanger
1983. 88 figures, 7 tables. XI, 348 pages
ISBN 3-540-12140-4

Springer-Verlag
Berlin
Heidelberg
New York
Tokyo